Applied Mathematical Sciences

Volume 79

Applied Mathematical Sciences

1. *John:* Partial Differential Equations, 4th ed.
2. *Sirovich:* Techniques of Asymptotic Analysis.
3. *Hale:* Theory of Functional Differential Equations, 2nd ed.
4. *Percus:* Combinatorial Methods.
5. *von Mises/Friedrichs:* Fluid Dynamics.
6. *Freiberger/Grenander:* A Short Course in Computational Probability and Statistics.
7. *Pipkin:* Lectures on Viscoelasticity Theory.
9. *Friedrichs:* Spectral Theory of Operators in Hilbert Space.
11. *Wolovich:* Linear Multivariable Systems.
12. *Berkovitz:* Optimal Control Theory.
13. *Bluman/Cole:* Similarity Methods for Differential Equations.
14. *Yoshizawa:* Stability Theory and the Existence of Periodic Solution and Almost Periodic Solutions.
15. *Braun:* Differential Equations and Their Applications, 3rd ed.
16. *Lefschetz:* Applications of Algebraic Topology.
17. *Collatz/Wetterling:* Optimization Problems.
18. *Grenander:* Pattern Synthesis: Lectures in Pattern Theory, Vol I.
20. *Driver:* Ordinary and Delay Differential Equations.
21. *Courant/Friedrichs:* Supersonic Flow and Shock Waves.
22. *Rouche/Habets/Laloy:* Stability Theory by Liapunov's Direct Method.
23. *Lamperti:* Stochastic Processes: A Survey of the Mathematical Theory.
24. *Grenander:* Pattern Analysis: Lectures in Pattern Theory, Vol. II.
25. *Davies:* Integral Transforms and Their Applications, 2nd ed.
26. *Kushner/Clark:* Stochastic Approximation Methods for Constrained and Unconstrained Systems
27. *de Boor:* A Practical Guide to Splines.
28. *Keilson:* Markov Chain Models—Rarity and Exponentiality.
29. *de Veubeke:* A Course in Elasticity.
30. *Sniatycki:* Geometric Quantization and Quantum Mechanics.
31. *Reid:* Sturmian Theory for Ordinary Differential Equations.
32. *Meis/Markowitz:* Numerical Solution of Partial Differential Equations.
33. *Grenander:* Regular Structures: Lectures in Pattern Theory, Vol. III.
34. *Kevorkian/Cole:* Perturbation methods in Applied Mathematics.
35. *Carr:* Applications of Centre Manifold Theory.
36. *Bengtsson/Ghil/Källén:* Dynamic Meteorology: Data Assimilation Methods.
37. *Saperstone:* Semidynamical Systems in Infinite Dimensional Spaces.
38. *Lichtenberg/Lieberman:* Regular and Stochastic Motion.
39. *Piccini/Stampacchia/Vidossich:* Ordinary Differential Equations in R^n.
40. *Naylor/Sell:* Linear Operator Theory in Engineering and Science.
41. *Sparrow:* The Lorenz Equations: Bifurcations, Chaos, and Strange Attractors.
42. *Guckenheimer/Holmes:* Nonlinear Oscillations, Dynamical Systems and Bifurcations of Vector Fields.
43. *Ockendon/Tayler:* Inviscid Fluid Flows.
44. *Pazy:* Semigroups of Linear Operators and Applications to Partial Differential Equations.
45. *Glashoff/Gustafson:* Linear Optimization and Approximation: An Introduction to the Theoretical Analysis and Numerical Treatment of Semi-Infinite Programs.
46. *Wilcox:* Scattering Theory for Diffraction Gratings.
47. *Hale et al.:* An Introduction to Infinite Dimensional Dynamical Systems—Geometric Theory.
48. *Murray:* Asymptotic Analysis.
49. *Ladyzhenskaya:* The Boundary-Value Problems of Mathematical Physics.
50. *Wilcox:* Sound Propagation in Stratified Fluids.
51. *Golubitsky/Schaeffer:* Bifurcation and Groups in Bifurcation Theory, Vol. I.
52. *Chipot:* Variational Inequalities and Flow in Porous Media.
53. *Majda:* Compressible Fluid Flow and Systems of Conservation Laws in Several Space Variables.
54. *Wasow:* Linear Turning Point Theory.

O. Hernández-Lerma

Adaptive Markov Control Processes

Springer-Verlag
New York Berlin Heidelberg
London Paris Tokyo

O. Hernández-Lerma
Departamento de Matemáticas
Centro de Investigación del IPN
07000 México, D.F.
México

With 2 Illustrations.

Mathematics Subject Classification (1980): 90C40, 93E20

Library of Congress Cataloging-in-Publication Data
Hernández-Lerma, O. (Onésimo)
Adaptive Markov control processes / O. Hernández-Lerma.
p. cm. — (Applied mathematical sciences ; v. 79)
Bibliography: p.
Includes indexes.
ISBN 0-387-96966-7
1. Markov processes. 2. Adaptive control systems. I. Title.
II. Series: Applied mathematical sciences (Springer-Verlag New York Inc.) ; v. 79.
QA1.A647 vol. 79
[QA274.7]
510 s—dc19
[519.2'33] 89-6015

Camera-ready copy prepared using LaTEX.
Printed and bound by R.R. Donnelley & Sons, Harrisonburg, Virginia.
Printed in the United States of America.

9 8 7 6 5 4 3 2 1

ISBN 0-387-96966-7 Springer-Verlag New York Berlin Heidelberg
ISBN 3-540-96966-7 Springer-Verlag Berlin Heidelberg New York

To Marina, Adrián, Gerardo,
and my father

Preface

This book is concerned with a class of discrete-time stochastic control processes known as *controlled Markov processes* (CMP's), also known as Markov decision processes or Markov dynamic programs. Starting in the mid-1950s with Richard Bellman, many contributions to CMP's have been made, and applications to engineering, statistics and operations research, among other areas, have also been developed. The purpose of this book is to present some recent developments on the theory of *adaptive* CMP's, i.e., CMP's that depend on *unknown* parameters. Thus at each decision time, the controller or decision-maker must *estimate* the true parameter values, and then *adapt* the control actions to the estimated values. We do not intend to describe all aspects of stochastic adaptive control; rather, the selection of material reflects our own research interests.

The prerequisite for this book is a knowledge of real analysis and probability theory at the level of, say, Ash (1972) or Royden (1968), but no previous knowledge of control or decision processes is required. The presentation, on the other hand, is meant to be self-contained, in the sense that whenever a result from analysis or probability is used, it is usually stated in full and references are supplied for further discussion, if necessary. Several appendices are provided for this purpose.

The material is divided into six chapters. Chapter 1 contains the basic definitions about the stochastic control problems we are interested in; a brief description of some applications is also provided. The main developments are presented in Chapter 2 for discounted-reward problems, and in Chapter 3 for average-reward problems. In each of these chapters we begin by presenting the basic theory and then we study several approximations to Markov control models. Finally the approximation results, which are important in themselves, are used to obtain adaptive control policies, i.e., policies combining parameter estimates and control actions. In Chapter 4 we consider partially observable control models and show how the adaptive control results in Chapters 2 and 3 can be extended to this situation. Chapter 5 introduces a method of parameter estimation in CMP's, and Chapter 6 presents discretization procedures for a class of adaptive and non-adaptive control problems. The chapter dependence is as follows:

$$\begin{array}{ccccc} 3 & & & & \\ \uparrow & & & & \\ 1 & \rightarrow & 2 & \rightarrow & 4 \\ \downarrow & & \downarrow & & \\ 5 & & 6 & & \end{array}$$

In bringing this book to its present form I have received help from many people. To begin, I should mention Steven I. Marcus for first kindling my interest in adaptive control problems. The book itself started as a set of lecture notes for short courses that I gave to graduate students in applied mathematics and electrical engineering at the Institute of Basic Sciences of the University of Veracruz (México, in May 1985 and May 1986), and at the Electrical Engineering Department of the University of Campinas (Brazil, in July 1985), respectively. Several parts of the book were also used in a graduate course on topics in Advanced Mathematical Statistics at Texas Tech University, Mathematics Department, during the academic year 1987/88. I want to thank Luis Cruz Kuri and Rafael Davis Velati (Veracruz), José C. Geromel (Campinas), and Ben Duran (Texas Tech) for their hospitality. Diego Bricio Hernández suggested corrections and other improvements on an earlier manuscript. Also my grateful thanks go to Roberto S. Acosta Abreu and Ben Duran for substantial comments on a previous draft and on the present version. I appreciate the support of the Sistema Nacional de Investigadores (SNI) and the Consejo Nacional de Ciencia y Tecnología (CONACYT). The latter institution has partially funded much of my research work in the last few years. Finally, I wish to thank Ms. Laura Valencia for typing several versions of the manuscript.

México City Onésimo Hernández-Lerma

Contents

Summary of Notation and Terminology

Abbreviations

a.s.	almost surely (i.e., with probability 1)
i.i.d.	independent and identically distributed
p.m.	probability measure
ADO	asymptotically discount optimal
CMP	controlled Markov process
CO	completely observable
DP(E)	dynamic programming (equation)
MCM	Markov control model
(N)VI	(nonstationary) value iteration
PEC	principle of estimation and control
PO	partially observable
SC	strongly consistent

Remark. The qualifying "a.s." is usually omitted in relationships involving conditional probabilities and expectations.

Symbols

$\mathbf{R}$	set of real numbers with the usual topology
$\Box$	end of proof
$:=$	equality by definition
1_B	indicator function of the set B defined as

$$1_B(x) := 1 \text{ if } x \in B, \quad \text{and} \quad := 0 \text{ if } x \notin B.$$

The following symbols are all defined in Section 1.2:

$\mathbf{K}$	set of admissible state-control pairs
$\mathbf{F}$	set of stationary policies
Δ	set of all admissible control policies
H_t	space of histories, h_t, up to time t
P_x^δ	p.m. when the policy $\delta \in \Delta$ is used and the initial state is $x_0 = x$
E_x^δ	expectation operator with respect to the p.m. P_x^δ

Spaces

A topological space X is *always* endowed with the Borel sigma-algebra $\mathcal{B}(X)$, i.e., the smallest sigma-algebra of subsets of X which contains all the open subsets of X. Thus "measurability" of sets and functions always means "Borel-measurability."

The Cartesian product of a (finite or countable) sequence of topological spaces $X_1, X_2, \ldots$ is denoted by $X_1 X_2 \ldots$, and is endowed with the product topology and the product sigma-algebra $\mathcal{B}(X_1)\mathcal{B}(X_2)\ldots$, i.e., the smallest sigma-algebra containing all finite-dimensional measurable rectangles $B_1 B_2 \ldots B_n$, where B_i is in $\mathcal{B}(X_i)$ for all $i = 1, \ldots, n$ and $n \geq 1$.

A *Borel space* is a Borel subset of a complete separable metric space. A Borel subset of a Borel space is itself a Borel space. (In particular, the empty set is regarded as a Borel space.) The following are examples of Borel spaces:

- $\mathbf{R}^n$ with the usual topology;
- a countable set X with the discrete topology (i.e., the topology consisting of all subsets of X);
- a compact metric space (which is complete and separable);
- if $X_1, X_2, \ldots$ is a (finite or countable) sequence of Borel spaces, then the product space $Y := X_1 X_2 \ldots$ is also a Borel space with Borel sigma-algebra

$$\mathcal{B}(Y) = \mathcal{B}(X_1)\mathcal{B}(X_2)\ldots.$$

Throughout the following, let X and Y be Borel spaces. We use the following notation:

$\mathbf{P}(X)$ space of p.m.'s on X with the topology of weak convergence (see Appendix B). Since X is a Borel space, $\mathbf{P}(X)$ is also a Borel space.

p_x p.m. on X concentrated at the point $x \in X$, i.e., $p_x(B) := 1_B(x)$ for $B \in \mathcal{B}(X)$.

$q(dy|x)$ a *stochastic kernel* (or conditional probability measure) on Y given X, i.e., a function such that, for each $x \in X$, $q(\cdot \mid x)$ is a p.m. on Y, and for each Borel set $B \in \mathcal{B}(Y)$, $q(B \mid \cdot)$ is a measurable function on X. (See Appendix C.)

$B(X)$ [respectively, $C(X)$] Banach space of real-valued bounded measurable (respectively, bounded continuous) functions v on X with the supremum norm $\|v\| := \sup_{x \in X} |v(x)|$.

$\|\mu\|$ *total variation* norm of a finite signed measure μ on X; see Appendix B.

1

Controlled Markov Processes

1.1 Introduction

The objective of this chapter is to introduce the stochastic control processes we are interested in; these are the so-called (discrete-time) *controlled Markov processes* (CMP's), also known as Markov decision processes or Markov dynamic programs. The main part is Section 1.2. It contains some basic definitions and the statement of the optimal and the adaptive control problems studied in this book. In Section 1.3 we present several examples; the idea is to illustrate the main concepts and provide sources for possible applications. Also in Section 1.3 we discuss (briefly) more general control systems, such as non-stationary CMP's and semi-Markov control models. The chapter is concluded in Section 1.4 with some comments on related references.

1.2 Stochastic Control Problems

The definition of an optimal control problem, for either deterministic or stochastic systems, requires three components:

- a decision or control model,
- a set of admissible control policies (or control strategies), and
- a performance index (or objective function).

These components are discussed in this section for the case of controlled Markov processes, and then the optimal (and the adaptive) control problem is stated.

Control Models

2.1 Definition. A discrete-time, stationary, *Markov control model* (or MCM for short) consists of four objects (X, A, q, r), where:

(a) X, the so-called *state space*, is a Borel space (i.e., a Borel subset of a complete separable metric space). The elements of X are called *states*.

(b) A, the *action* (or *control*) *set*, is a Borel space. To each state $x \in X$ we associate a non-empty measurable subset $A(x)$ of A, whose elements are the *admissible actions* (or controls) when the system is in state x. The set $\mathbf{K}$ of admissible state-action pairs, defined by

$$\mathbf{K} := \{(x,a) \mid x \in X \text{ and } a \in A(x)\},$$

is assumed to be a measurable subset of the product space XA. The elements (x,a) of $\mathbf{K}$ are sometimes denoted by k.

(c) q, or $q(B \mid k)$, where $B \in \mathcal{B}(X)$ and $k \in \mathbf{K}$, is the *transition law*, a stochastic kernel on X given $\mathbf{K}$.

(d) $r : \mathbf{K} \to \mathbf{R}$ is a measurable function and is called the *one-step reward* (or return or revenue) *function*.

The MCM (X, A, q, r) in Definition 2.1 is interpreted as representing a controlled stochastic system which is observed at times $t = 0, 1, \ldots$; the state and control at time t are denoted by x_t and a_t, respectively, and the system evolves as follows. If the system is in state $x_t = x$ at time t and the control action $a_t = a \in A(x)$ is chosen, then we receive a reward $r(x,a)$ and the system moves to a new state x_{t+1} according to the probability distribution $q(\cdot \mid x, a)$ on X, i.e.,

$$q(B \mid x,a) = \text{Prob}(x_{t+1} \in B \mid x_t = x, a_t = a) \quad \text{for all } B \in \mathcal{B}(X).$$

Once the transition into the new state, say, $x_{t+1} = x'$, has occurred, a new control $a' \in A(x')$ is chosen and the process is repeated.

In most applications the state space X is either $\mathbf{R}^n$ or a denumerable (possibly finite) set. There are situations, however, which require a more general setting. For instance, in the "partially observable" systems studied in Chapter 4 we need to consider control problems with the state space being a set of probability measures. All these cases are included in Definition 2.1 by allowing X to be a Borel space.

Some authors write a MCM as $(X, A, \{A(x), x \in X\}, q, r)$, including the sets $A(x)$ in the description, but we will use the shorter notation (X, A, q, r) introduced in Definition 2.1. Continuous-time and non-stationary MCM's will be briefly discussed in Section 1.3 below.

In *adaptive* control problems we shall consider MCM's $(X, A, q(\theta), r(\theta))$ in which the transition law $q = q(\theta)$ and the one-step reward function $r = r(\theta)$ depend measurably on a parameter θ whose "true" value, say, θ^*, is *unknown;* we do know, though, that the "admissible" parameter values lie in a given parameter set Θ, a Borel space.

Throughout the following we consider a fixed MCM (X, A, q, r).

Policies

For each $t \geq 0$, we define the space H_t of (admissible) ***histories*** up to time t (or t-histories) by $H_0 := X$ and

$$H_t := \mathbf{K}^t X = \mathbf{K} H_{t-1} \quad \text{if } t \geq 1.$$

A generic element h_t of H_t is a vector of the form

2.2 $$h_t = (x_0, a_0, \ldots, x_{t-1}, a_{t-1}, x_t)$$

where $(x_i, a_i) \in \mathbf{K}$ for all $i = 0, \ldots, t-1$, and $x_t \in X$.

2.3 Definition.

(a) A (randomized, admissible) ***policy*** is a sequence $\delta = \{\delta_t\}$ of stochastic kernels δ_t on A given H_t satisfying the constraint

$$\delta_t(A(x_t) \,|\, h_t) = 1 \quad \text{for all} \quad h_t \in H_t \quad \text{and} \quad t \geq 0.$$

We denote by Δ the set of all policies.

(b) A ***deterministic policy*** is a sequence $\{f_t\}$ of measurable functions f_t from H_t to A such that

$$f_t(h_t) \in A(x_t) \quad \text{for all} \quad h_t \in H_t \quad \text{and} \quad t \geq 0.$$

A deterministic policy $\{f_t\}$ can be regarded as a (randomized) policy $\delta = \{\delta_t\}$ such that, for each $h_t \in H_t$, $\delta_t(\cdot \,|\, h_t)$ is the probability measure on A, concentrated at $f_t(h_t)$, that is,

$$\delta_t(B \,|\, h_t) = 1_B[f_t(h_t)] \quad \text{for all} \quad B \in \mathcal{B}(A), \;\; h_t \in H_t, \;\; \text{and } t \geq 0,$$

where 1_B is the indicator function of the set B (see the Summary of Notation and Terminology).

The (randomized and deterministic) policies in Definition 2.3 are also known as "history-dependent" policies and they are particularly useful in adaptive control problems that depend on unknown parameters, since, to *estimate* those parameters at a given time t, one usually needs to know the history (h_t) of the system up to time t. However, in standard (non-adaptive) control problems sometimes it suffices to consider policies which depend only on the current state, say x_t, and not on the whole history h_t. These are the so-called Markov (or feedback, or closed-loop, or memoryless) policies defined as follows.

2.4 Definition. Let $\mathbf{F}$ be the set of all ***decision functions*** (or selectors), i.e., measurable functions $f : X \to A$ such that $f(x) \in A(x)$ for all $x \in X$. Equivalently, $\mathbf{F}$ is the product space $\Pi_{x \in X} A(x)$.

(a) A *Markov policy* is a sequence $\{f_t\}$ of functions $f_t \in \mathbf{F}$ for all t; thus the control applied at time t is $a_t := f_t(x_t)$.

(b) A Markov policy $\{f_t\}$ such that $f_t = f$ is independent of t is called a *stationary* policy. We shall refer to a stationary policy $\{f, f, \ldots\}$ simply as the stationary policy $f \in \mathbf{F}$; in other words, we shall identify $\mathbf{F}$ with the set of stationary policies. (Some authors denote a stationary policy $\{f, f, \ldots\}$ by f^∞.)

Clearly, a Markov policy can (and will) be regarded as a deterministic policy in the sense of Definition 2.3(b), and therefore, we have the following relations

$$\mathbf{F} \subset \{\text{Markov policies}\} \subset \{\text{deterministic policies}\} \subset \Delta.$$

2.5 Remarks on the Underlying Probability Space and the Markov Property. Let $(\Omega, \mathcal{F})$ be the measurable space in which Ω is the product space $(XA)^\infty = XAXA\ldots$ and $\mathcal{F}$ is the corresponding product sigma-algebra. The elements of Ω are infinite sequences of the form

$$\omega = (x_0, a_0, x_1, a_1, \ldots), \text{ where } x_t \in X \text{ and } a_t \in A \text{ for all } t \geq 0,$$

and the state and control variables x_t and a_t are defined as projections (or coordinate variables) from Ω to the sets X and A, respectively. Note that Ω contains the space $H_\infty = \mathbf{K}\,\mathbf{K}\cdots$ of (admissible) histories

$$(x_0, a_0, x_1, a_1 \ldots) \text{ with } (x_t, a_t) \in \mathbf{K} \text{ for all } t \geq 0.$$

Moreover, by a theorem of C. Ionescu Tulcea (Proposition C.3 in Appendix C), for any given policy $\delta = \{\delta_t\} \in \Delta$ and any initial state $x_0 = x \in X$, there exists a unique probability measure P_x^δ on $(\Omega, \mathcal{F})$ given by

$$\begin{aligned} &P_x^\delta(dx_0, da_0, dx_1, da_1, \ldots) \\ &= p_x(dx_0)\delta_0(da_0 \mid x_0) q(dx_1 \mid x_0, a_0)\delta_1(da_1 \mid x_0, a_0, x_1) \ldots, \end{aligned}$$

where p_x is the probability measure on X concentrated at $\{x\}$, and satisfying:

2.5 (a) $P_x^\delta(H_\infty) = 1$,
2.5 (b) $P_x^\delta(x_0 = x) = 1$,
2.5 (c) $P_x^\delta(a_t \in B \mid h_t) = \delta_t(B \mid h_t)$ for all $B \in \mathcal{B}(A)$, $h_t \in H_t$ and $t \geq 0$, and
2.5 (d) $P_x^\delta(x_{t+1} \in C \mid h_t, a_t) = q(C \mid x_t, a_t)$ for all $C \in \mathcal{B}(X)$, $h_t \in H_t$ and $t \geq 0$.

(If the initial state x_0 is random with distribution, say, q_0, we replace p_x above by q_0.)

The stochastic process $(\Omega, \mathcal{F}, P_x^\delta, \{x_t\})$ is said to be a *controlled Markov process* (CMP).

The expectation operator with respect to the probability measure P_x^δ is denoted by E_x^δ.

Equation 2.5(d) is called the Markov property and it can be stated in the more general form 2.5(e) below [see also 2.5(g)], for which we introduce some notation. For any $t \geq 1$ let us write a t-history $h_t \in H_t$ as $h_t = (i_t, x_t)$, where

$$i_t := (x_0, a_0, \ldots, x_{t-1}, a_{t-1}) \in \mathbf{K}^t,$$

so that if $h_n \in H_n$ is some n-history, say,

$$h_n = (x'_0, a'_0, \ldots, x'_{n-1}, a'_{n-1}, x'_n), \text{ where } n \geq 0,$$

then (i_t, h_n) is a $(t+n)$-history in H_{t+n}. Now, given any policy $\delta = \{\delta_t\}$ let us define the "t-shifted" policy $\delta^{(t)} = \{\delta_n^{(t)},\ n = 0, 1, \ldots\}$ by

$$\delta_n^{(t)}(\cdot \mid h_n) := \delta_{t+n}(\cdot \mid (i_t, h_n)) \text{ for all } n \geq 0 \text{ and } h_n \in H_n.$$

That is,

$$\delta_0^{(t)}(\cdot \mid x'_0) = \delta_t(\cdot \mid (i_t, x'_0)) \text{ for all } x'_0 \in H_0,$$

$$\delta_1^{(t)}(\cdot \mid h_1) = \delta_{t+1}(\cdot \mid (i_t, h_1)) \text{ for all } h_1 = (x'_0, a'_0, x'_1) \in H_1,$$

etc. Note that $\delta^{(0)} = \delta$, and if $f \in \mathbf{F}$ is a *stationary* policy, then $f^{(t)} = f$ for all $t \geq 0$. We can now state the following.

2.5 (e) For any measurable set $D \in \mathcal{F}$ and any $t \geq 0$,

$$P_x^\delta\{(a_t, x_{t+1}, a_{t+1}, x_{t+2}, \ldots) \in D \mid h_t = (i_t, x_t)\}$$
$$= P_{x_t}^{\delta^{(t)}}\{(a'_0, x'_1, a'_1, x'_2, \ldots) \in D\}.$$

For instance, using 2.5(c) and 2.5(d) above, a direct calculation shows that for any two Borel sets $B \in \mathcal{B}(A)$ and $C \in \mathcal{B}(X)$ and any $t \geq 0$,

$$P_x^\delta(a_t \in B, x_{t+1} \in C \mid h_t) = P_{x_t}^{\delta^{(t)}}(a'_0 \in B, x'_1 \in C).$$

Indeed, the left-hand side (l.h.s.) can be written [by definition of P_x^δ and 2.5(c) and 2.5(d)] as

$$P_x^\delta(a_t \in B, x_{t+1} \in C \mid h_t) = \int_X \delta_t(da_t \mid h_t) q(C \mid x_t, a_t),$$

whereas the right-hand side (r.h.s.) is given by

$$P_{x_t}^{\delta^{(t)}}(a'_0 \in B, x'_1 \in C) = \int_X p_{x_t}(dx'_0) \int_B \delta_0^{(t)}(da'_0 \mid x'_0) q(C \mid x'_0, a'_0)$$
$$= \int_B \delta_t(da'_0 \mid (i_t, x_t)) q(C \mid x_t, a'_0),$$

and the equality follows.

Property 2.5(e) is equivalent to the following.

2.5 (f) For any P_x^δ-integrable function b and any $t \geq 0$,

$$E_x^\delta[b(a_t, x_{t+1}, a_{t+1}, x_{t+2}, \ldots) \mid h_t] = E_{x_t}^{\delta^{(t)}}[b(a_0', x_1', a_1', x_2', \ldots)].$$

Finally, using standard properties of conditional probabilities it can be verified that if $\delta = \{f_t\}$ is a *Markov* policy, so that in 2.5(c) we have

$$\delta_t(B \mid h_t) = 1_B[f_t(x_t)] \quad \text{for all} \quad B \in \mathcal{B}(A),\ h_t \in H_t \quad \text{and} \quad t \geq 0,$$

then the state process $\{x_t\}$ is a Markov process in the usual sense, that is,

2.5 (g) $P_x^\delta(x_{t+1} \in C \mid x_0, x_1, \ldots, x_t) = P_x^\delta(x_{t+1} \in C \mid x_t) = q(C \mid x_t, f_t(x_t))$

for all $C \in \mathcal{B}(X)$ and $t \geq 0$. Indeed, if $\delta = \{f_t\}$ is a Markov policy, then

$$P_x^\delta(x_{t+1} \in C \mid x_0, \ldots, x_t) = E_x^\delta\{P_x^\delta(x_{t+1} \in C \mid h_t) \mid x_0, \ldots, x_t\}$$

[from 2.5(c) and 2.5(d)]

$$\begin{aligned} &= E_x^\delta\{q(C \mid x_t, f_t(x_t)) \mid x_0, \ldots, x_t\} \\ &= q(C \mid x_t, f_t(x_t)), \end{aligned}$$

and a similar argument gives

$$P_x^\delta(x_{t+1} \in C \mid x_t) = q(C \mid x_t, f_t(x_t)).$$

This implies 2.5(g). In particular, for a *stationary* policy $f \in \mathbf{F}$, $\{x_t\}$ is a time-homogeneous Markov process with transition kernel

$$P_x^f(x_{t+1} \in C \mid x_t) = q(C \mid x_t, f(x_t)) \quad \text{for all} \quad C \in \mathcal{B}(X) \quad \text{and} \quad t \geq 0.$$

In *adaptive* MCM's $(X, A, q(\theta), r(\theta))$, the probability and expectation P_x^δ and E_x^δ above will be written as $P_x^{\delta,\theta}$ and $E_x^{\delta,\theta}$, respectively, to emphasize the dependence on θ.

Performance Criteria

Once we have a (Markov) control model (X, A, q, r) and a set Δ of admissible policies, to complete the description of an optimal control problem we need to specify a *performance index* (or *objective function*), that is, a function "measuring" the system's performance when a given policy $\delta \in \Delta$ is used and the initial state of the system is x.

Here we are concerned (mainly) with two performance criteria. The first one (to be studied in Chapter 2) is the *expected total discounted reward*

2.6
$$V(\delta, x) := E_x^\delta\left[\sum_{t=0}^{\infty} \beta^t r(x_t, a_t)\right],$$

where $\delta \in \Delta$ is the policy used and $x_0 = x \in X$ is the initial state. The number β in 2.6 is called the *discount factor* and to insure that $V(\delta, x)$ is a finite (actually, bounded) function for all δ and x, it will be assumed that the one-step reward $r(x, a)$ is bounded and $\beta \in (0, 1)$. In some applications it is convenient to write β as $1/(1+\alpha)$, where $\alpha > 0$ is an "interest rate"; see, e.g., Clark (1976), p. 69. [On the other hand, there are problems in which it is convenient to consider $\beta \geq 1$, but the treatment of this situation is technically different from the so-called *discounted case* above when $\beta \in (0, 1)$ and $r(x, a)$ is bounded; see, e.g., Hernández-Lerma and Lasserre (1988), Hübner (1977), or Rieder (1979).]

The second performance index (studied in Chapter 3) is the long-run average expected reward per unit time, or simply the *average reward* criterion defined by

2.7
$$J(\delta, x) := \liminf_{n \to \infty} n^{-1} E_x^\delta \left[\sum_{t=0}^{n-1} r(x_t, a_t) \right],$$

where $\delta \in \Delta$, $x_0 = x \in X$, and again, r is a bounded function. Taking the lim inf (and not just the limit, which might not exist) insures that $J(\delta, x)$ is a well-defined finite-valued function for all $\delta \in \Delta$ and $x \in X$. [We could take instead the lim sup, but in general the end results are not quite the same; the lim inf criterion is somewhat easier to deal with! See, e.g., Cavazos-Cadena (1988; 88b) or Flynn (1976).]

Control Problems

Finally, suppose we are given a MCM (X, A, q, r), a set Δ of admissible policies, and a performance index, say (to fix ideas), the discounted reward $V(\delta, x)$ in 2.6. The *optimal control problem* is then defined as follows: Determine a policy $\delta^* \in \Delta$ such that

2.8
$$V(\delta^*, x) = \sup_{\delta \in \Delta} V(\delta, x) \text{ for all } x \in X.$$

Any policy satisfying 2.8 (if one exists!) is said to be *optimal*. The function v^* defined by

$$v^*(x) := \sup_{\delta \in \Delta} V(\delta, x), \quad \text{for } x \in X,$$

is called the *optimal reward* (or optimal value) *function*. (We have implicitly assumed above that Δ is non-empty; this will always be the case for the control problems in the subsequent chapters.)

In *adaptive* control problems, the MCM $(X, A, q(\theta), r(\theta))$ — and therefore, the performance index

$$V(\delta, x, \theta) := E_x^{\delta, \theta} \left[\sum_{t=0}^{\infty} \beta^t r(x_t, a_t, \theta) \right]$$

and the optimal reward function

$$v^*(x,\theta) := \sup_{\delta\in\Delta} V(\delta,x,\theta) \text{ — },$$

depend on unknown parameters θ, so the problem is then to find an *adaptive* optimal policy δ^*_θ, that is, a policy combining parameter estimation and control actions, and such that

2.9 $$V(\delta^*_\theta,x,\theta) = v^*(x,\theta) \quad \text{for all} \quad x \in X,$$

when θ is the true parameter value.

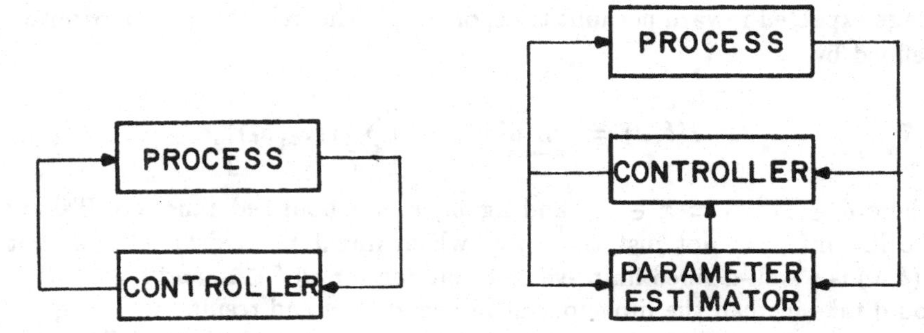

Figure 1. A standard (feedback) control system.

Figure 2. An adaptive control system.

The adaptive control problem and the standard optimal control problem are substantially different; they are schematically illustrated in Figures 1 and 2. In the standard case, roughly, the controller observes the state x_t of the system and then (possibly using additional information from the history h_t) chooses a control, say, $a_t = a_t(h_t)$. In the adaptive case, however, before choosing a_t, the controller gets an estimate θ_t of the unknown parameter value and combines this with the history h_t to select a control action of the form $a_t(h_t,\theta_t)$. On the other hand, all statistical results giving convergence of the estimates to the true parameter value are *asymptotic* results (except, perhaps, in trivial cases), and therefore, the only way we can get general optimality conditions such as 2.9 is "in the limit", as $t \to \infty$. This is the main reason for using (in adaptive control problems) mainly "infinite horizon" performance criteria, as those in 2.6 and 2.7. Actually, in some cases, the notion of optimality as defined by 2.9 turns out to be too strong and it has to be replaced by a weaker notion (an example of this is the concept of asymptotic discount optimality introduced in Chapter 2).

One final remark. To estimate the unknown parameters it might be necessary to use "auxiliary observations" $z_0, z_1, \ldots$, so that instead of the histories h_t in 2.2 one would use "extended" vectors of the form

$$h'_t := (x_0, z_0, a_0, \ldots, x_{t-1}, z_{t-1}, a_{t-1}, x_t, z_t)$$

to compute the controls a_t at time t. This approach requires some minor changes but it leads to the same final results, and therefore, to simplify the exposition, we will work with stochastic control problems as defined in the previous paragraphs. [This is a standard convention in adaptive control; see, e.g., Georgin (1978b), Kumar (1985), Mandl (1974),... .] For details on the "extended" setting the reader is refered to Kurano (1985), Schäl (1981), Van Hee (1978), Kolonko (1982a).

We will now consider some examples to illustrate the concepts introduced in this section.

1.3 Examples

Perhaps the simplest example of a controlled Markov process is a discrete-time *deterministic* (i.e., non-random) control system

3.1 $x_{t+1} = F(x_t, a_t)$, with $t = 0, 1, \ldots$, (and $x_t \in X$, $a_t \in A(x_t)$, and so on),

where $F : \mathbf{K} \to X$ is a given measurable function, and the initial state x_0 is some given point in X. In terms of Definition 2.1, we can also describe the system 3.1 by a MCM (X, A, q, r) with transition law $q(B \mid k)$ given by

$$q(B \mid k) = 1_B[F(k)] \quad \text{for all } B \in \mathcal{B}(X) \text{ and } k = (x, a) \in \mathbf{K};$$

i.e., $q(\cdot \mid k)$ is the p.m. on X concentrated at $F(k)$ for all k in $\mathbf{K}$. The admissible policies in the present case are the deterministic policies in Definition 2.3(b). [However, for technical reasons it is sometimes convenient to use randomized policies, or "relaxed controls", as in 2.3(a); see, e.g., Arstein (1978) and references therein. Of course, if one uses a randomized policy, the control system is no longer "deterministic".]

Another class of examples of controlled Markov processes consists of control systems defined by a "system equation" such as

3.2 $$x_{t+1} = F(x_t, a_t, \xi_t), \quad \text{with } t = 0, 1, \ldots,$$

where $\{\xi_t\}$, the so-called *disturbance* or *driving* process, is a sequence of independent and identically distributed (i.i.d.) random elements with values in some Borel space S, and common probability distribution μ. The initial state x_0 is assumed to be independent of $\{\xi_t\}$, and F is a measurable function from $\mathbf{K}S$ to X. Again, expressed in terms of Definition 2.1, we can write system 3.2 as a MCM (X, A, q, r) in which the transition law q, i.e.,

$$q(B \mid k) = \text{Prob}(x_{t+1} \in B \mid x_t = x, \ a_t = a),$$

is given by

3.3 $q(B \mid k) = \int_S 1_B[F(k, s)]\mu(ds)$ for all $B \in \mathcal{B}(X)$ and $k = (x, a) \in \mathbf{K}$, or equivalently, $q(B \mid k) = \mu(\{s \in S \mid F(k,s) \in B\})$. In Chapter 2 we will consider (among others) the adaptive control problem in which the unknown "parameter" is the distribution μ.

In many applied control systems, the state space X is a *countable* set (with the discrete topology, in which case X is a Borel space). This situation occurs in control of queueing systems, quality control, machine maintenance problems, etc.; see, e.g., Heyman and Sobel (1984), Ross (1970), Dynkin and Yushkevich (1979),... . This case (X countable) is somewhat simpler than the general case (X a Borel space), because one can then resort to many results in elementary Markov chain theory. For adaptive control problems with countable X, references will be provided in the appropriate places.

We will now discuss briefly some specific control problems. Their solutions are not given here (references are provided); the idea is simply to illustrate the concepts introduced in Section 1.2 and provide sources for possible applications.

An Inventory/Production System

Consider a finite capacity ($C < \infty$) inventory/production system in which the state variable x_t is the stock level at the beginning of period t, where $t = 0, 1, \ldots$. The control variable a_t is the quantity ordered or produced (and immediately supplied) at the beginning of period t, and the "disturbance" process $\{\xi_t\}$ is the demand, a sequence of i.i.d. random variables with distribution μ. Denoting the amount sold during period t by

$$y_t := \min(\xi_t, x_t + a_t),$$

the system equation becomes

3.4 $$x_{t+1} = x_t + a_t - y_t = (x_t + a_t - \xi_t)^+,$$

where $v^+ := \max(0, v)$ and the initial state is some given inventory level x_0 independent of $\{\xi_t\}$.

The state space and the control set are $X = A = [0, C]$, whereas the set of admissible controls in state x is $A(x) = [0, C - x]$. Suppose the demand distribution μ is absolutely continuous with density m, i.e.,

$$\mu(D) = \int_D m(s)ds \quad \text{for all} \quad D \in \mathcal{B}(\mathbf{R}).$$

Then, from 3.3, the transition law $q(B \mid k)$, for any Borel subset B of X and any admissible state-action pair $k = (x, a) \in \mathbf{K}$, becomes

3.5 $$q(B \mid x, a) = \int 1_B[(x + a - s)^+]m(s)ds.$$

Thus if $B = \{0\}$, then

$$q(\{0\} \mid x, a) = \int_{x+a}^{\infty} m(s)ds,$$

and if B is contained in $(0, C]$, then

$$q(B \mid x, a) = \int_0^{x+a} 1_B(s) m(x + a - s) ds.$$

The (expected) one-stage reward $r(x, a)$ may have different forms, depending on the specific situation we have in mind. For instance, if we are given the unit sale price (p), the unit production cost (c), and a unit holding cost (h), all positive numbers, then the net revenue at stage t is

$$r_t = py_t - ca_t - h(x_t + a_t),$$

and $r(x, a)$ becomes

3.6
$$\begin{aligned} r(x,a) &= E(r_t \mid x_t = x, a_t = a) \\ &= \int [p \cdot \min(s, x+a) - ca - h(x+a)] m(s) ds. \end{aligned}$$

This completes the specification of the MCM (X, A, q, r) as in Definition 2.1, and of the set Δ of admissible policies (Definition 2.3), since these are determined by the control constraint sets $A(x)$. Finally, an optimal control problem would be specified by giving a performance index, such as (e.g.) 2.6 or 2.7.

Inventory problems have been studied extensively in the stochastic control/operations research literature [Bensoussan (1982), Bertsekas (1976; 1987), DeGroot (1970), Kushner (1971), Ross (1970), etc.]. In an *adaptive* inventory control problem, the demand distribution μ might depend on an unknown parameter [Georgin (1978b)], or the distribution itself might be unknown. [For adaptive inventory problems following the Bayesian approach, see, e.g., Waldmann (1980, 1984) and references therein.]

Control of Water Reservoirs

An important source of (deterministic and stochastic) control problems are those related to water reservoir operations. An excellent introduction to many of these problems, including the connections between these and inventory systems, is given by Yakowitz (1982).

In a simplified situation, the system equation for a water reservoir with finite capacity C is

3.7
$$x_{t+1} = \min(x_t - a_t + \xi_t, C), \text{ where } t = 0, 1, \ldots,$$

and x_t, the state variable, is the volume—or stock—of water at time t. The control a_t is the volume of water released (for irrigation or to produce electrical energy, say) during period t, and the "disturbance" ξ_t is the water inflow during that period. Assuming the disturbance process is a sequence of i.i.d. random variables, this problem could be stated in MCM-form, as the inventory control problem above. However, instead of doing this, we simply remark that water reservoir control systems are typical examples of systems with partial state information, or ***partially observable*** (PO). That is, as a rule, we do not know directly what the water stock x_t is. Usually x_t is ***estimated*** using observations or measurements (e.g., the water ***level***) of the form

3.8 $$y_t = G(a_{t-1}, x_t, \eta_t)$$

where $\{\eta_t\}$ is a stochastic process representing the measurements errors. We thus have a PO control system of the type to be discussed in Chapter 4. Other examples of PO systems are learning or artificial intelligence processes, and statistical hypothesis testing.

Fisheries Management

In recent years there has been a growing concern for problems in (renewable and non-renewable) resource management and control of biological populations, e.g., forest management [Clark (1976), Lembersky (1978)], epidemics and pest control [Jaquette (1972), Lefevre (1981), Wickwire (1977)], and oil exploration [Andreatta and Runggaldier (1986)]. In particular, fisheries management problems pose special difficulties which result from environmental and interspecific influences and the complexity of related economic aspects [Clark (1976), Palm (1977), Walters (1978), Walters and Hilborn (1976), Lovejoy (1984), Ludwig and Walters (1982)]. Leaving aside a few exceptions [e.g., Getz and Swartzman (1981)], the dynamic models developed for fisheries systems are of the general form

$$x_{t+1} = F(x_t, a_t, \xi_t),$$

where x_t is the vector of state variables, such as fish population size and current level of economic development; a_t is the vector of control actions, such as harvest rates; and ξ_t is a vector of "disturbance" variables: random environmental effects, intrinsic growth processes and so on. In the simplest case, a single variable x_t representing population size (in numbers or biomass) is taken as the main biological determinant of the state of the system, and a typical dynamic model is, e.g., the Ricker model

$$x_{t+1} = (x_t - a_t)\exp[\alpha - \beta(x_t - a_t) + \xi_t],$$

where α and β are positive parameters. The complexity of fisheries systems makes them a natural source of adaptive control problems in which, being "economic" problems, the performance index is usually the discounted reward criterion 2.6.

To simplify the exposition, in the examples presented above we have considered only MCM's in which the transition law is determined by an explicit system equation such as 3.2, 3.4, or 3.7. There are many situations, however, in which this is not the case, as in surveillance [Yakowitz et al. (1976)] and random search problems [Hinderer (1970)], learning models [Monahan (1982), Bertsekas (1987)], quality control and machine maintenance problems [Monahan (1982), Wang (1976), C.C. White (1978)], or fishing systems [Getz and Swartzman (1981)], just to name a few. In these examples, the "natural" dynamic model is provided by a transition law q (typically a transition matrix), and not by a "system function" F as in 3.2, which might be too complicated to specify. The two approaches are supposed to be "mathematically" equivalent [Yushkevich and Chitashvili (1982), p. 243], but for "practical" purposes we can easily go from 3.2 to compute q, as in 3.3, but not backwards, where "backwards" means: Given a MCM (X, A, q, r), find a function F and a sequence $\{\xi_t\}$ of i.i.d. random elements with some common distribution μ such that 3.2 and 3.3 hold. Thus, to provide a general setting, we prefer to use a transition law q as done in Definition 2.1.

There are many other applications of controlled Markov processes; a long list has been collected by D.J. White (1985). On the other hand, there are stochastic control models which are not MCM's in the sense of Definition 2.1, but still, they can be reduced to the form (X, A, q, r). Some of these control models are the following.

Nonstationary MCM's

The MCM (X, A, q, r) in Definition 2.1 is *stationary*, which means that the defining data, namely, X, A, q and r, does not vary from stage to stage. In contrast, a *non-stationary* MCM is of the form $(X_t,\ A_t,\ q_t,\ r_t,\ t \in \mathbf{N})$, where $\mathbf{N} = \{0, 1, \ldots\}$, and X_t and A_t are Borel spaces denoting, respectively, the state space and the control set at time t. If $x_t = x \in X_t$, the set of admissible actions is then $A_t(x)$, a non-empty measurable subset of A_t; if in addition $a_t = a \in A_t(x)$, then $q_t(\cdot \mid x, a)$ is the distribution of x_{t+1}, given $x_t = x$ and $a_t = a$, and $r_t(x, a)$ is the expected reward for period t.

A non-stationary MCM can be reformulated in stationary form (X, A, q, r) by a "state augmentation" procedure, as follows. Define

$$X := \{(x, t) \mid x \in X_t, t \in \mathbf{N}\},$$

$$A := \{(a, t) \mid a \in A_t, t \in \mathbf{N}\},$$

and the set of admissible controls in state $(x, t) \in X$ by

$$A[(x, t)] := A_t(x) \cdot \{t\} = \{(a, t) \mid a \in A_t(x)\}.$$

Finally, for any admissible state-action pair $((x, t), (a, t))$, define the one-step reward as

$$r((x, t), (a, t)) := r_t(x, a),$$

and the transition law $q(\cdot \mid (x, t), (a, t))$ as a probability measure that assigns probability one to $X_{t+1} \cdot \{t+1\}$ with marginal $q_t(\cdot \mid x, a)$ in the first variable, i.e.,

$$q(B \cdot \{t+1\} \mid (x,t),(a,t)) = q_t(B \mid x,a) \quad \text{for all} \quad B \in \mathcal{B}(X_{t+1}),$$

where $B \cdot \{t+1\} = \{(b,t+1) \mid b \in B\}$.

A similar reduction can be done for more general non-stationary models. For instance, Schäl (1975), Section 8, considers discounted rewards (see 2.6) with time- and state-dependent discount factors $\beta_t(x_t, a_t, x_{t+1})$, and one-step rewards $r_t(x_t, a_t, x_{t+1})$, whereas Bertsekas and Shreve (1978), Section 10.1, consider transition laws $q_t(\cdot \mid x, a)$ determined by dynamic models (cf. 3.2)

$$x_{t+1} = F_t(x_t, a_t, \xi_t).$$

where $\{\xi_t\}$ is a sequence of random elements $\xi_t \in S_t$ with distribution of the form $\mu_t(\cdot \mid x_t, a_t)$ depending on the state and control at time t; that is,

$$q_t(B \mid x,a) = \mu_t(\{s \in S_t \mid F_t(x,a,s) \in B\} \mid x,a)$$

for all $B \in \mathcal{B}(X_{t+1})$, $x \in X_t$ and $a \in A_t(x)$. In addition to the cited references, non-stationary models are studied (e.g.) by Hinderer (1970) and Striebel (1975).

Semi-Markov Control Models

In the "discrete-time" MCM (X, A, q, r) of Definition 2.1 the "decision epochs" are the fixed times $t = 0, 1, \ldots$. However, there are control problems in which the *decision epochs* are *random* times $0 = \sigma_0 \leq \sigma_1 \leq \ldots$. Consider, for instance, the problem of control of service in a queueing system [e.g., Hernández-Lerma and Marcus (1983)]. Roughly, there are jobs to be processed, one at a time, in a "service station"; since service is not instantaneous, arriving jobs have to wait in line—the "queue"—while a job is being processed. When a service is completed, the server (i.e., the controller) has to decide the rate "a" at which he will process the next job. Thus the decision epochs, starting at $\sigma_0 = 0$, are the "service completion times", and the *interdecision times*

$$\tau_n := \sigma_n - \sigma_{n-1}, \quad \text{where} \quad n = 1, 2, \ldots,$$

are the (random) service times. In this example, the state of the system is the number $x(t)$ of jobs (waiting in line or in service) in the system at time $t \in T$, where $T := [0, \infty)$, and the state at the nth decision epoch is $x_n := x(\sigma_n)$. A general setting for this type of problems is provided by the (Markov renewal or) semi-Markov control models (SMCM's) defined as follows.

3.9 Definition. A SMCM is given by (X, A, q, G, r), where X, A, q and r (and also $A(x)$ and $\mathbf{K}$) are as in Definition 2.1 and $G(dt \mid x, a, y)$ is a stochastic kernel on $T = [0, \infty)$ given $\mathbf{K}X$.

The idea is the following: if at any given decision epoch, σ_n, the state of the system is $x_n = x(\sigma_n) = x$ in X, and a control $a \in A(x)$ is applied, then a reward $r(x,a)$ is immediately received and the time until the next decision epoch σ_{n+1} and the corresponding state $x_{n+1} = x(\sigma_{n+1})$ have joint probability distribution

$$Q(BC \mid x,a) := \text{Prob}(x_{n+1} \in B, \tau_{n+1} \in C \mid x_n = x, a_n = a)$$

given by

3.10
$$Q(BC \mid x,a) = \int_B G(C \mid x,a,y) q(dy \mid x,a)$$

for all $B \in \mathcal{B}(X)$, $C \in \mathcal{B}(T)$ and $(x,a) \in \mathbf{K}$, where $\tau_{n+1} = \sigma_{n+1} - \sigma_n$. Thus $G(\cdot \mid x,a,y)$ is interpreted as the conditional distribution of τ_{n+1} given $x_n = x$, $a_n = a$ and $x_{n+1} = y$.

Two important cases occur when G does *not* depend on y, that is, $G(\cdot \mid x,a,y) = G(\cdot \mid x,a)$, so that, from 3.10,

$$Q(BC \mid x,a) = G(C \mid x,a) q(B \mid x,a).$$

The first one is when $G(\cdot \mid x,a)$ is an exponential distribution, say

$$G([0,t] \mid x,a) = 1 - \exp[-\lambda(x,a)t] \quad \text{if } t \geq 0,$$

and zero otherwise, where $\lambda(x,a)$ is a positive measurable function on $\mathbf{K}$. In this case, the SMCM is said to be a *continuous-time* MCM. The second case is when $\tau_n \equiv 1$ for all n, so that $G(\cdot \mid x,a)$ is concentrated at $t = 1$ for all $(x,a) \in \mathbf{K}$, and the SMCM reduces to the *discrete-time* MCM of Definition 2.1.

More generally, there are elementary transformations to reduce (under appropriate conditions) a continuous-time SMCM to an *equivalent* discrete-time MCM. This has been done by many authors; see, e.g., Federgruen and Tijms (1978), Kakumanu (1977), Mine and Tabata (1970), Morton (1973), Schweitzer (1971), Serfozo (1979), etc.

1.4 Further Comments

The study of controlled Markov processes (or Markov decision processes) began in the early 1950's and the first systematic treatment was done by Bellman (1957). He also introduced the term "adaptive" control of Markov chains in Bellman (1961). In addition to the Bellman books, there are presently many introductory texts dealing with CMP's, e.g., Bertsekas (1976) and (1987), Heyman and Sobel (1984), Kumar and Varaiya (1986), Mine and Osaki (1970), Ross (1970; 1983), etc. Somewhat more advanced books are those by Dynkin and Yushkevich (1979), Bertsekas and Shreve

(1978), and Hinderer (1970). Our presentation in Section 1.2 (at least for non-adaptive control problems) is partly a summary of related concepts in the last two references. For applications, see the references provided in Section 1.3 above.

2

Discounted Reward Criterion

2.1 Introduction

Let (X, A, q, r) be a Markov control model (MCM) as defined in Section 1.2 (Definition 2.1). We consider in this chapter the problem of maximizing the expected total discounted reward defined as

1.1 $$V(\delta, x) := E_x^\delta \sum_{t=0}^{\infty} \beta^t r(x_t, a_t) \text{ for } \delta \in \Delta \text{ and } x \in X,$$

where $\beta \in (0,1)$ is the discount factor. Sometimes we also write $V(\delta, x)$ as $V_\delta(x)$. The *optimal reward function* is defined by

1.2 $$v^*(x) := \sup_{\delta \in \Delta} V(\delta, x) \text{ for } x \in X.$$

Under the assumptions given in Section 2.2, v^* is a real-valued measurable function on the state space X.

We will consider two notions of optimality: one is the standard concept of discount optimality, while the other one is an *asymptotic* definition introduced by Schäl (1981) to study *adaptive* control problems in the discounted case.

1.3 Definition. A policy δ is called

(a) *discount optimal* (DO) if $V(\delta, x) = v^*(x)$ for every $x \in X$.

(b) *asymptotically discount optimal* (ADO) if, for every $x \in X$,

$$|V_n(\delta, x) - E_x^\delta v^*(x_n)| \to 0 \quad \text{as} \quad n \to \infty,$$

where

$$V_n(\delta, x) := E_x^\delta \sum_{t=n}^{\infty} \beta^{t-n} r(x_t, a_t)$$

is the expected total discounted reward from stage n onward.

Summary

We begin in Section 2.2 by giving conditions under which v^* satisfies the Dynamic Programming Equation (Theorem 2.2), and in Section 2.3 we relate asymptotic discount optimality to a function that measures the "discrepancy" between an optimal action in state x and any other action

$a \in A(x)$. (Recall that $A(x)$ denotes the set of admissible controls in state x; see Section 1.2.)

In Section 2.4 we present a nonstationary value-iteration (NVI) procedure to approximate dynamic programs and to obtain ADO policies. To illustrate the NVI procedure, we also give in Section 2.4 a finite-state approximation scheme for denumerable state controlled Markov processes.

In Sections 2.5 and 2.6 we study *adaptive* control problems. That is, we consider MCM's $(X, A, q(\theta), r(\theta))$ with transition law $q(\theta)$ and one-step reward function $r(\theta)$ depending on an unknown parameter θ. In particular, in Section 2.6, the unknown parameter is the distribution of the (i.i.d.) disturbance process $\{\xi_t\}$ in a discrete-time system

$$x_{t+1} = F(x_t, a_t, \xi_t).$$

The NVI scheme(s) of Section 2.4 are used to give a unified presentation of several adaptive policies.

We conclude in Section 2.7 with some comments on the results obtained and on the related literature.

In the proofs we use some results from analysis which are collected as appendices.

2.2 Optimality Conditions

Recall that, by definition of a MCM (Section 1.2), the state space X and the action set A are Borel spaces, and the set

$$\mathbf{K} := \{(x, a) \mid x \in X \text{ and } a \in A(x)\}$$

is assumed to be a measurable subset of XA. *Throughout this chapter* we assume that, in addition, the MCM (X, A, q, r) satisfies the following.

2.1 Assumptions.

(a) For each state $x \in X$, the set $A(x)$ of admissible controls is a (non-empty) compact subset of A.

(b) For some constant R, $|r(k)| \leq R$ for all $k = (x, a) \in \mathbf{K}$, and moreover, for each x in X, $r(x, a)$ is a continuous function of $a \in A(x)$.

(c) $\int v(y)\, q(dy \mid x, a)$ is a continuous function of $a \in A(x)$ for each $x \in X$ and each function $v \in B(X)$.

In 2.1(c), $B(X)$ is the Banach space of real-valued bounded measurable functions on X with the supremum norm $\|v\| := \sup_x |v(x)|$. (See the Summary of Notation and Terminology.) Note that, by Assumption 2.1(b), the reward functions are uniformly bounded: $|V(\delta, x)| \leq R/(1-\beta)$ for every policy δ and initial state x.

Remark. All the results in this section hold if in 2.1(b) and (c) we replace "continuous" by "upper semi-continuous"; see, e.g., Himmelberg et al. (1976).

The main objective of this section is to prove the following.

2.2 Theorem. *Under Assumptions* 2.1,

(a) *The optimal reward function v^* is the unique solution in $B(X)$ of the (discounted-reward) dynamic programming equation (DPE)*

$$v^*(x) = \max_{a \in A(x)} \{r(x,a) + \beta \int_X v^*(y)\, q(dy \mid x,a)\} \quad \textit{for} \quad x \in X.$$

The DPE is also known as the (discounted-reward) optimality equation.

(b) *A stationary policy $f^* \in \mathbf{F}$ is optimal if and only if $f^*(x)$ maximizes the right-hand side (r.h.s.) of the DPE for all $x \in X$, that is,*

$$v^*(x) = r(x, f^*(x)) + \beta \int v^*(y)\, q(dy \mid x, f^*(x)) \quad \textit{for all} \quad x \in X. \tag{1}$$

2.3 Remark. Assumptions 2.1 insure the existence of a stationary policy $f^* \in \mathbf{F}$ satisfying equation (1). Indeed, under those assumptions, the part in brackets in the DPE, namely,

$$r(x,a) + \beta \int v^*(y)\, q(dy \mid x,a),$$

is a measurable function in $(x,a) \in \mathbf{K}$, and *continuous* in $a \in A(x)$ for all $x \in X$, where $A(x)$ is a *compact* set. Thus the existence of such an $f^* \in \mathbf{F}$ follows from Proposition D.3 in Appendix D. This kind of argument to show the existence of "measurable selectors" $f^* \in \mathbf{F}$ will be used repeatedly, sometimes without explicit reference to Appendix D.

To prove Theorem 2.2 we need some preliminary results.

Let T be the operator on $B(X)$ defined by

2.4
$$Tv(x) := \max_{a \in A(x)} \left\{ r(x,a) + \beta \int_X v(y)\, q(dy \mid x,a) \right\}$$

for all $v \in B(X)$ and $x \in X$. We call T the *dynamic programming* (DP) *operator.* Using Proposition D.3 in Appendix D (as above, in Remark 2.3), it can be seen that $Tv \in B(X)$ whenever $v \in B(X)$. Note also that the DPE can be written as

$$v^* = Tv^*.$$

We define another operator T_g on $B(X)$, for each stationary policy $g \in \mathbf{F}$, by

$$T_g v(x) := r(x, g(x)) + \beta \int_X v(y)\, q(dy \mid x, g(x)),$$

where $v \in B(X)$ and $x \in X$. Note that equation (1) in Theorem 2.2(b) becomes

$$v^* = T_{f^*} v^*.$$

2.5 Lemma. *Both T and T_g, for every $g \in \mathbf{F}$, are contraction operators with modulus β; therefore, by Banach's Fixed Point Theorem (Proposition A.1 in Appendix A), there exists a unique function $u^* \in B(X)$ and a unique function $v_g \in B(X)$ such that*

$$Tu^* = u^* \quad \textit{and} \quad T_g v_g = v_g,$$

and moreover, for any function $v \in B(X)$,

$$\|T^n v - u^*\| \to 0 \quad \textit{and} \quad \|T_g^n v - v_g\| \to 0 \quad \textit{as } n \to \infty.$$

Proof. To prove that T is a contraction operator we can use either Proposition A.2 or Proposition A.3 (in Appendix A). Using the latter, it follows that for any two functions v and u in $B(X)$ and any $x \in X$.

$$|Tv(x) - Tu(x)| \leq \max_{a \in A(x)} \beta \left| \int [v(y) - u(y)]\, q(dy \,|\, x, a) \right| \leq \beta \|u - v\|,$$

and then, taking the sup over all $x \in X$, we obtain $\|Tv - Tu\| \leq \beta\|v - u\|$. The result for T_g is obvious. □

We want now to relate the "fixed points" u^* and v_g in Lemma 2.5 to the optimal reward function v^* and the reward $V(g, x)$ when using the stationary policy g. We begin in the following lemma by showing that $v_g = V_g$, where we write $V(g, x)$ as $V_g(x)$.

2.6 Lemma.

(a) *$v_g = V_g$ for every stationary policy $g \in \mathbf{F}$.*

(b) *A policy δ^* is optimal if and only if its reward satisfies the DPE, i.e., $TV(\delta^*, x) = V(\delta^*, x)$ for every $x \in X$.*

Proof. (a) By the uniqueness of the fixed point v_g of T_g (Lemma 2.5), it suffices to verify that V_g satisfies the equation $V_g = T_g V_g$. To do this, let us expand $V_g(x)$ as follows:

$$V_g(x) = E_x^g \sum_{t=0}^{\infty} \beta^t\, r(x_t, a_t) = r(x, g(x)) + \beta E_x^g \left[\sum_{t=1}^{\infty} \beta^{t-1}\, r_t \right],$$

where we have written $r_t := r(x_t, a_t)$. By standard properties of conditional expectations and the Markov property [Remark 2.5(f) in Chapter 1], the expectation on the r.h.s. becomes

$$\begin{aligned}
E_x^g\left[\sum_1^\infty \beta^{t-1} r_t\right] &= E_x^g\left[E_x^g\left(\sum_1^\infty \beta^{t-1} r_t \,|\, h_1\right)\right] \\
&= E_x^g\left[E_{x_1}^g\left(\sum_1^\infty \beta^{t-1} r_t\right)\right] \\
&= E_x^g[V_g(x_1)] \\
&= \int V_g(y)\, q(dy \,|\, x, g(x)),
\end{aligned}$$

so that

$$V_g(x) = r(x, g(x)) + \beta \int V_g(y)\, q(dy \,|\, x, g(x)) = T_g V_g(x) \quad \text{for every} \quad x \in X.$$

(b) First, we prove the "if" part. Thus let δ^* be a policy such that its reward $u(x) := V(\delta^*, x)$ satisfies the DPE $u = Tu$:

$$u(x) = \max_{a \in A(x)} \left\{ r(x, a) + \beta \int u(y)\, q(dy \,|\, x, a) \right\}, \quad x \in X.$$

We will show that δ^* is optimal, i.e., $u(x) \geq V(\delta, x)$ for every policy δ, and every initial state $x \in X$. To simplify the notation, in this proof we fix an arbitrary policy $\delta \in \Delta$ and $x \in X$, and write E_x^δ simply as E.

Now for any history $h_t \in H_t$, it follows from the Markov property 2.5(d) in Section 1.2 that

$$\begin{aligned}
E[\beta^{t+1} u(x_{t+1}) \,|\, h_t, a_t] &= \beta^{t+1} \int u(y)\, q(dy \,|\, x_t, a_t) \\
&= \beta^t \left\{ r(x_t, a_t) + \beta \int u(y)\, q(dy \,|\, x_t, a_t) \right\} \\
&\quad - \beta^t r(x_t, a_t) \leq \beta^t u(x_t) - \beta^t r(x_t, a_t),
\end{aligned}$$

or equivalently,

$$\beta^t u(x_t) - E[\beta^{t+1} u(x_{t+1}) \,|\, h_t, a_t] \geq \beta^t r(x_t, a_t).$$

Therefore, taking expectations $E = E_x^\delta$ and summing over $t = 0, \ldots, n$, we obtain

$$u(x) - \beta^{n+1} E\, u(x_{n+1}) \geq E \sum_{t=0}^n \beta^t r(x_t, a_t).$$

Finally, letting $n \to \infty$ we get the desired conclusion, $u(x) \geq V(\delta, x)$; that is, δ^* is optimal.

To prove the converse (the "only if" part), let us assume that δ^* is optimal. We will show that $u(x) := V(\delta^*, x)$ satisfies (i) $u \leq Tu$, and (ii)

$u \geq Tu$, so that u satisfies the DPE. To prove (i) we expand $u(x)$ as in the proof of part (a) above to obtain

$$\begin{aligned} u(x) &= E_x^{\delta^*} \sum_{t=0}^{\infty} \beta^t r(x_t, a_t) \\ &= \int_A \delta_0^*(da \mid x) \left\{ r(x,a) + \beta \int_X V[\delta^{*(1)}, y]\, q(dy \mid x, a) \right\}, \end{aligned}$$

where $\delta^{*(1)} = \{\delta_t^{*(1)}\}$ denotes the "1-shifted" policy in Section 1.2, Remark 2.5; that is, with $x_0 = x$ and $a_0 = a$,

$$\delta_t^{*(1)}(\cdot \mid h_t) := \delta_{t+1}^*(\cdot \mid x_0, a_0, h_t) \quad \text{for} \quad t = 0, 1, \ldots .$$

Thus since (by assumption) δ^* is optimal,

$$\begin{aligned} u(x) &\leq \int_A \delta_0^*(da \mid x) \left\{ r(x,a) + \beta \int_X u(y)\, q(dy \mid x, a) \right\} \\ &\leq \max_{a \in A(x)} \left\{ r(x,a) + \beta \int u(y)\, q(dy \mid x, a) \right\} \\ &= Tu(x), \end{aligned}$$

which proves (i).

To prove inequality (ii), let $g \in \mathbf{F}$ be an arbitrary stationary policy, and let $\delta' := (g, \delta^*)$ be the policy that uses g at time $t = 0$, and uses the optimal policy δ^* from time $t = 1$ onwards, i.e., $\delta_0'(x_0) := g(x_0)$, and for $t \geq 1$,

$$\delta_t'(\cdot \mid x_0, a_0, \ldots, x_{t-1}, a_{t-1}, x_t) := \delta_{t-1}^*(\cdot \mid x_1, a_1, \ldots, x_t).$$

Thus the optimality of δ^* implies

$$u(x) \geq V(\delta', x) = r(x, g(x)) + \beta \int u(y)\, q(dy \mid x, g(x)) \quad \text{for all} \quad x \in X,$$

so that, since $g \in \mathbf{F}$ is arbitrary,

$$u(x) \geq \max_{a \in A(x)} \left\{ r(x,a) + \beta \int u(y)\, q(dy \mid x, a) \right\} = Tu(x).$$

This completes the proof of Lemma 2.6. □

We can now obtain Theorem 2.2 from Lemmas 2.5 and 2.6.

Proof of Theorem 2.2. (a) By Lemma 2.6(b), a policy δ^* is optimal, that is, $V(\delta^*, x) = v^*(x)$ for all $x \in X$, if and only if v^* satisfies the DPE $v^* = Tv^*$, and the uniqueness of such a solution (or fixed point) v^* follows from Lemma 2.5.

(b) Suppose $f \in \mathbf{F}$ is a stationary policy satisfying equation (1), i.e., $v^* = T_f v^*$. Then Lemmas 2.5 and 2.6(a) imply

$$v^* = v_f = V_f,$$

and therefore, f is optimal. Conversely, if $f \in \mathbf{F}$ is optimal, then $v^* = v_f$ and the uniqueness of the fixed point v_f implies $v^* = T_f v^*$, i.e., f satisfies (1). This completes the proof of Theorem 2.2. □

2.7 Remark. *Value-iteration.* Let $v_n := T^n v = T v_{n-1}$ be the functions defined in Lemma 2.5; that is, $v_0 := v \in B(X)$ is arbitrary and

$$v_n(x) := \max_{a \in A(x)} \left\{ r(x,a) + \beta \int v_{n-1}(y)\, q(dy \mid x, a) \right\}$$

for all $n \geq 1$ and $x \in X$. Note that, by the contraction property of T,

$$\|v_n - v^*\| = \|T v_{n-1} - T v^*\| \leq \beta \, \|v_{n-1} - v^*\|,$$

so that $\|v_n - v^*\| \leq \beta^n \|v_0 - v^*\|$ for all $n \geq 0$. The v_n are called the *value-iteration* (or successive approximation) functions, and we will use them in later sections to obtain asymptotically discount optimal policies.

Continuity of v^*

There are some situations (e.g., in Section 2.6) in which it is required to have a *continuous* optimal reward function v^*. This, of course, requires in general more restrictive assumptions than 2.1 above.

2.8 Theorem. *Suppose:*

(a) *For each $x \in X$, the set $A(x)$ is compact and, moreover, the set-valued mapping $x \longrightarrow A(x)$ is continuous.* (See Appendix D.)

(b) $r \in C(\mathbf{K})$, *where, for any topological space S, $C(S)$ denotes the Banach space of real-valued bounded continuous functions on S endowed with the* sup norm.

(c) *The transition law $q(\cdot \mid k)$ is (weakly-) continuous on* $\mathbf{K}$, i.e.,

$$\int v(y)\, q(dy \mid k)$$

is a continuous function of $k = (x, a) \in \mathbf{K}$ for every $v \in C(X)$.

Then the optimal reward function v^* is the unique solution in $C(X)$ of the DPE.

The proof is the same as that of Theorem 2.2; in particular, it uses again that the DP operator T, defined now on $C(X)$, is a contraction operator.

In Section 2.6 we will give conditions under which v^* is Lipschitz-continuous.

2.3 Asymptotic Discount Optimality

If a policy δ is discount optimal (DO), then it is asymptotically discount optimal (ADO) in the sense of Definition 1.3(b), i.e., for every $x \in X$,

3.1 $$|V_n(\delta, x) - E_x^\delta v^*(x_n)| \to 0 \quad \text{as} \quad n \to \infty.$$

This results from Bellman's Principle of Optimality in Hinderer (1970, p. 109; or p. 19 when X is countable), namely: δ is optimal, i.e., $v^*(\cdot) = V(\delta, \cdot)$, if and only if

$$E_x^\delta v^*(x_n) = V_n(\delta, x) \quad \text{for every} \quad n \geq 0 \quad \text{and} \quad x \in X,$$

in which case, the left-hand side (l.h.s.) of 3.1 is zero.

The reason for introducing the (weaker) asymptotic definition is that for *adaptive* MCM's $(X, A, q(\theta), r(\theta))$, there is no way one can get optimal policies, in general, because of the errors introduced when computing the reward

$$V(\delta, x, \theta) := E_x^{\delta,\theta} \left[\sum_{t=0}^{\infty} \beta^t r(x_t, a_t, \theta) \right]$$

with the "estimates" $\hat{\theta}_t$ of the true (but unknown) parameter value θ. Thus the idea behind 3.1 is to allow the system to run during a "learning period" of n stages, and then we compare the reward V_n, discounted from state n onwards, with the expected optimal reward when the system's "initial state" is x_n. The ADO concept was introduced by Schäl (1981). (The situation is different for *average* reward problems; for these, it is possible to obtain *optimal*—instead of asymptotically optimal—adaptive policies. This is discussed in Section 3.1.)

Our objective in this section is to characterize asymptotic discount optimality in terms of the function ϕ from $\mathbf{K}$ to $\mathbf{R}$ defined by

3.2 $$\phi(x, a) := r(x, a) + \beta \int v^*(y)\, q(dy \mid x, a) - v^*(x).$$

This function was first used in (average-reward) adaptive control problems by Mandl (1974), but it also appears in other contexts as a measure of the "discrepancy" between an optimal action in state x and any other action $a \in A(x)$; see Remark 3.7 below. Note that $\phi \leq 0$ and, under Assumption 2.1, the following properties are immediate.

3.3 Proposition.

(a) *ϕ is a bounded measurable function on* $\mathbf{K}$.

(b) *For each $x \in X$, $\phi(x, a)$ is a continuous function of $a \in A(x)$.*

Moreover, the optimality Theorem 2.2 can be rewritten in terms of ϕ as follows.

3.4 Proposition.

(a) *DPE:* $\sup_{a\in A(x)} \phi(x,a) = 0$.

(b) *Optimality criterion: A stationary policy $f \in \mathbf{F}$ is (discount) optimal if and only if $\phi(x, f(x)) = 0$ for all $x \in X$.*

Now, to relate $\phi(x,a)$ to the concept of ADO, first note that

$$\phi(x_t, a_t) = E_x^\delta\{r(x_t, a_t) + \beta v^*(x_{t+1}) - v^*(x_t) \mid h_t, a_t\}$$

for any $x \in X$, $\delta \in \Delta$, and $t \geq 0$. Next, multiply by β^{t-n}, take expectation E_x^δ, and sum over all $t \geq n$, to obtain

3.5 $$\sum_{t=n}^{\infty} \beta^{t-n} E_x^\delta \phi(x_t, a_t) = V_n(\delta, x) - E_x^\delta v^*(x_n).$$

Comparing this with the l.h.s. of 3.1 we see that δ is ADO if and only if, for every $x \in X$,

$$\sum_{t=n}^{\infty} \beta^{t-n} E_x^\delta \phi(x_t, a_t) \to 0 \quad \text{as} \quad n \to \infty,$$

which is clearly equivalent to: for every $x \in X$,

$$E_x^\delta \phi(x_t, a_t) \to 0 \quad \text{as} \quad t \to \infty.$$

This in turn implies that

$$\phi(x_t, a_t) \to 0 \ \text{ in probabilty-}P_x^\delta \ \text{ for every } \ x \in X.$$

The converse is also true, since ϕ is bounded (Proposition 3.3) and therefore, uniformly integrable [Ash (1972), Theorem 7.5.2, p. 295]; or, alternatively, we could prove the converse using the extension of the Dominated Convergence Theorem in Ash (1972), p. 96. We summarize this discussion as follows.

3.6 Theorem. *A policy δ is ADO if and only if $\phi(x_t, a_t) \to 0$ in probability-P_x^δ for every $x \in X$.*

3.7 Remark. [Cf. Cavazos-Cadena (1986), Section 5.] Let $f \in \mathbf{F}$ and $\delta' \in \Delta$ be arbitrary, and let $\delta := \{f, \delta'\}$ be the policy that uses f at time $t = 0$, and then uses δ' from time 1 onward; that is, $\delta_0(x_0) := f(x_0)$, and for $t \geq 1$ and every history $h_t = (x_0, a_0, \ldots, x_{t-1}, a_{t-1}, x_t)$,

$$\delta_t(\cdot \mid h_t) := \delta'_{t-1}(\cdot \mid x_1, a_1, \ldots, x_t).$$

Then, for every $x \in X$,

$$\begin{aligned} V(\delta, x) &= r(x, f(x)) + \beta \int V(\delta', y)\, q(dy \mid x, f(x)) \\ &\leq r(x, f(x)) + \beta \int v^*(y)\, q(dy \mid x, f(x)) \\ &= v^*(x) + \phi(x, f(x)), \end{aligned}$$

or, since $|\phi| = -\phi$, $v^*(x) - V(\delta, x) \geq |\phi(x, f(x))|$. Thus, since $f \in \mathbf{F}$ is arbitrary, we conclude that

$$v^*(x) - V(\delta, x) \geq |\phi(x, a)|$$

for any policy $\delta = \{\delta_t\}$ whose initial action is $\delta_0(x) = a \in A(x)$ when $x_0 = x$. This means that we can interpret $|\phi(x, a)|$ as the smallest "deviation from optimality" we can hope for when using the control action a in state x.

In the following sections Theorem 3.6 will be used to obtain approximating and adaptive ADO policies. In particular, in Section 2.4 we will use the following definition suggested by Theorem 3.6.

3.8 Definition. A Markov policy $\{f_t\}$, i.e., a sequence of functions $f_t \in \mathbf{F}$ is said to be

(a) asymptotically discount optimal (ADO) if, for every $x \in X$,

$$\phi(x, f_t(x)) \to 0 \quad \text{as} \quad t \to \infty, \text{ and}$$

(b) uniformly ADO if

$$\sup_{x \in X} |\phi(x, f_t(x))| \to 0 \quad \text{as} \quad t \to \infty.$$

As an example of an uniformly ADO Markov policy, let $\delta = \{f_t\}$ be such that $f_0 \in \mathbf{F}$ is arbitrary, and for $t \geq 1$, $f_t \in \mathbf{F}$ maximizes the r.h.s. of the value iteration (VI) functions in Remark 2.7, i.e.,

$$v_t(x) = r(x, f_t(x)) + \beta \int v_{t-1}(y)\, q(dy \mid x, f_t(x)) \tag{3.9}$$

for every $x \in X$, and $t \geq 1$. Then adding and subtracting $v_t(x)$ on the r.h.s. of

$$\phi(x, f_t(x)) = r(x, f_t(x)) + \beta \int v^*(y)\, q(dy \mid x, f_t(x)) - v^*(x),$$

a direct calculation show that

$$\sup_x |\phi(x, f_t(x))| \leq \beta \, \|v_{t-1} - v^*\| + \|v_t - v^*\|$$

(by Remark 2.7)

$$\leq 2\beta^t \|v_0 - v^*\| \to 0 \quad \text{as} \quad t \to \infty. \tag{3.10}$$

Thus $\delta = \{f_t\}$ is uniformly ADO for any initial function $v_0 \in B(X)$. Notice also that we can use equation 3.5 to estimate the "deviation from optimality" when using the VI policy δ from time n onward; i.e., from 3.5 and 3.10,

$$\begin{aligned} |V_n(\delta, x) - E_x^\delta v^*(x_n)| &= \sum_{t=n}^{\infty} \beta^{t-n} E_x^\delta \phi(x_t, f_t(x_t)) \\ &\leq 2 \, \|v_0 - v^*\| \, \beta^n / (1 - \beta). \end{aligned}$$

We will use this approach to obtain other ADO policies below.

2.4 Approximation of MCM's

A stochastic control problem can be approximated in many different ways. For instance, it can be approximated by finite-horizon problems, or by discretization procedures of the state space and/or the control set, as those to be introduced in Chapter 6.

Our approach in this section, however, is motivated by its applicability to adaptive MCM's $(X, A, q(\theta), r(\theta))$ with an unknown parameter θ. In this problems, the general idea is to take "estimates" θ_t of θ, and show that if θ_t "converges" to θ, then suitable (adaptive) policies defined in terms of the approximating sequence $(X, A, q(\theta_t), r(\theta_t))$, are ADO for the θ-model. To formalize these ideas, we first develop the procedure for general approximating models. We introduce three versions, NVI-1,2,3, of a nonstationary value iteration (NVI) scheme and corresponding Markov policies which are shown to be ADO for the limiting control model. In Sections 2.5 and 2.6 we identify the ADO policies with ***adaptive*** strategies. The results in this section were originally inspired by the NVI schemes introduced by Federgruen and Schweitzer (1981) for Markov decision problems with finite state and control spaces.

Nonstationary Value-Iteration

Let (X, A, q_t, r_t), where $t = 0, 1, \ldots$, be a sequence of MCM's each of which satisfies Assumptions 2.1 [in particular, $|r_t(k)| \leq R$ for all $k \in \mathbf{K}$ and $t \geq 0$], and such that they "converge" to the MCM (X, A, q, r) in the following sense.

4.1 Assumption. Both $\rho(t)$ and $\pi(t)$ converge to zero as $t \to \infty$, where

$$\rho(t) := \sup_k |r_t(k) - r(k)|$$

and

$$\pi(t) := \sup_k \|q_t(\cdot \mid k) - q(\cdot \mid k)\|.$$

Here the sup is over all $k \in \mathbf{K}$, and in the definition of $\pi(t)$, $\| \ \|$ denotes the total variation norm for finite signed measures (Appendix B).

4.2 Remark. Assumption 4.1 is equivalent to: as $t \to \infty$,

$$\bar{\rho}(t) := \sup_{s \geq t} \rho(s) \to 0 \quad \text{and} \quad \bar{\pi}(t) := \sup_{s \geq t} \pi(s) \to 0.$$

Observe also that both sequences $\bar{\rho}(t)$ and $\bar{\pi}(t)$ are non-increasing.

Now, for each $t \geq 0$, we define an operator T_t on $B(X)$ by

$$T_t v(x) := \max_{a \in A(x)} G_t(x, a, v), \tag{4.3}$$

where, for every $k = (x, a)$ in $\mathbf{K}$ and $v \in B(X)$,

$$G_t(k, v) := r_t(k) + \beta \int v(y)\, q_t(dy \mid k).$$

Note that T_t is the DP operator for the t-MCM (X, A, q_t, r_t); in other words, T_t is the same as the DP operator T in Section 2.2 above, with q and r replaced by q_t and r_t, respectively, which justifies the name of "nonstationary" value-iteration; see Remark 2.7. We shall exploit this relationship between T and T_t to obtain different uniform approximations of the optimal reward function v^* in Theorem 2.2 and ADO Markov policies—see Theorems 4.8 and 4.9 below.

To begin with, notice that for every $t \geq 0$, T_t is a contraction operator with modulus β, i.e.,

$$\|T_t u - T_t v\| \leq \beta \|u - v\| \quad \text{for all } u \text{ and } v \text{ in } B(X),$$

and then we can use Banach's Fixed Point Theorem (Appendix A) and Proposition D.3 in Appendix D to define the following.

NVI-1. For each $t \geq 0$, let $v_t^* \in B(X)$ be the unique fixed point of T_t, i.e.,

$$v_t^*(x) = T_t v_t^*(x) \quad \text{for all } x \in X, \tag{4.4}$$

(this is the DPE for the t-MCM), and let $\delta^* = \{f_t^*\}$ be a sequence of decision functions $f_t^* \in \mathbf{F}$ such that $f_t^*(x)$ maximizes the r.h.s. of 4.4 for every $x \in X$, i.e.,

$$v_t^*(x) = G_t(x, f_t^*(x), v_t^*).$$

(In other words, for each $t \geq 0$, $f_t^* \in \mathbf{F}$ is an optimal stationary policy for the t-MCM; see Theorem 2.2.)

NVI-2. For each $t \geq 0$, define $\overline{v}_t \in B(X)$ recursively:

$$\overline{v}_t(x) := T_t \overline{v}_{t-1}(x) = max_{a \in A(x)} G_t(x, a, \overline{v}_{t-1}) \text{ for all } x \in X, \tag{4.5}$$

where $\overline{v}_{-1}(\cdot) := 0$, and let $\overline{\delta} = \{\overline{f}_t\}$ be a sequence of decision functions such that $\overline{f}_t(x)$ maximizes the right side of 4.5 for every $x \in X$, i.e.,

$$\overline{v}_t(x) = G_t(x, \overline{f}_t(x), \overline{v}_{t-1}).$$

NVI-3. Let $\{w_t\}$ be any given sequence of functions in $B(X)$ satisfying

$$\|w_t - v_t^*\| \to 0 \quad \text{as} \quad t \to \infty, \tag{4.6}$$

where v_t^* are the functions in 4.4. Now, let $\{\epsilon_t\}$ be a decreasing sequence of positive numbers converging to zero, and for each $x \in X$, define the set of "ϵ_t-maximizers"

$$A_t(x) := \{a \in A(x) \mid G_t(x, a, w_t) \geq T_t w_t(x) - \epsilon_t\}.$$

Finally, let $\delta' = \{f'_t\}$ be a sequence of measurable functions from X to A such that $f'_t(x) \in A_t(x)$ for every $x \in X$ and $t \geq 0$. [As an example of how one may choose the functions w_t in 4.6, let $w_0 \in B(X)$ be such that $\|w_0\| \leq R$, and then define $w_t := T_t^{j(t)} w_{t-1}$ for $t = 1, 2, \ldots$, where $\{j(t)\}$ is a sequence of positive integers increasing to infinity. Then the contraction property of T_t yields $\|w_t - v_t^*\| \leq 2c_0\beta^{j(t)}$ for all $t \geq 1$ and c_0 as defined in 4.7 below.]

It turns out that, under Assumptions 4.1, each of the sequences in 4.4–4.6 converges uniformly to v^*, the optimal value function of the limiting MCM (X, A, q, r). To state this precisely, let us introduce the constants

4.7 $\quad c_0 := R/(1-\beta), \; c_1 := (1+\beta c_0)/(1-\beta), \;$ and $\; c_2 := c_1 + 2c_0.$

4.8 Theorem. *Suppose that Assumptions* 4.1 *hold. Then, for each* $t \geq 0$,

(a) $\|v_t^* - v^*\| \leq c_1 \cdot \max\{\rho(t), \pi(t)\}$.

(b) $\|\overline{v}_t - v^*\| \leq c_2 \cdot \max\{\overline{\rho}([t/2]), \overline{\pi}([t/2]), \beta^{[t/2]}\}$, *where* $[c]$ *denotes the largest integer less than or equal to c. Moreover, if the sequences* $\rho(t)$ *and* $\pi(t)$ *in* 4.1 *are non-increasing, then on the right side of* (b) *we can substitute* $\overline{\rho}$ *and* $\overline{\pi}$ *by* ρ *and* π, *respectively.*

(c) $\|w_t - v^*\| \leq \|w_t - v_t^*\| + \|v_t^* - v^*\| \to 0$.

We also have the following.

4.9 Theorem. *Under Assumptions* 4.1, *each of the Markov policies* δ^*, $\overline{\delta}$ *and* δ' *is uniformly ADO (Definition* 3.8) *for the MCM* (X, A, q, r).

Proof of Theorem 4.8. First note that part (c) follows from (a) and 4.6. Note also that the constant c_0 in 4.7 is an upper bound for v^*, v_t^* and $\overline{v}_t$:

$$\|v^*\| \leq c_0, \quad \|v_t^*\| \leq R + \beta\|v_t^*\|, \quad \text{and} \quad \|\overline{v}_t\| \leq R\sum_{s=0}^{t}\beta^s \leq c_0$$

for all $t \geq 0$. Let us now prove parts (a) and (b).

(a) From 4.4, the DPE in Theorem 2.2, and Proposition A.3 in Appendix A, we obtain, for every $x \in X$,

$$|v_t^*(x) - v^*(x)| \leq \max_{a \in A(x)} |r_t(x,a) - r(x,a) + \beta \int v_t^*(y)\, q_t(dy \mid x,a)$$
$$- \beta \int v^*(y)\, q(dy \mid x,a)|.$$

Inside the absolute value on the r.h.s., add and subtract the term

$$\beta \int v_t^*(y)\, q(dy \mid x,a);$$

then use the triangle inequality and the definitions of $\rho(t)$ and $\pi(t)$ to obtain

$$|v_t^*(x) - v^*(x)| \leq \rho(t) + \beta \|v_t^*\| \, \pi(t) + \beta \|v_t^* - v^*\|,$$

where we have used inequality B.1 in Appendix B. Finally, taking sup over all $x \in X$ and using that $\|v_t^*\| \leq c_0$, the above inequality becomes

$$(1-\beta)\|v_t^* - v^*\| \leq \rho(t) + \beta c_0 \pi(t) \leq (1+\beta c_0) \max\{\rho(t), \pi(t)\},$$

and (a) follows.

(b) [Cf. Federgruen and Schweitzer (1981).] A similar argument, using 4.5 and Theorem 2.2 again, yields

$$\|\overline{v}_{t+1} - v^*\| \leq \rho(t+1) + \beta \|v^*\| \, \pi(t+1) + \beta \|\overline{v}_t - v^*\|.$$

Therefore, for every $m = 1, 2, \ldots$,

$$\|\overline{v}_{t+m} - v^*\| \leq \sum_{k=0}^{m-1} \beta^k [\rho(t+m-k) + \beta c_0 \, \pi(t+m-k)] + \beta^m \|\overline{v}_t - v^*\|.$$

Now, since $\|v_t - v^*\| \leq 2c_0$, and $\overline{\rho}(t) \geq \rho(t+s)$ and $\overline{\pi}(t) \geq \pi(t+s)$ for all s, it follows from the previous inequality that

$$\|\overline{v}_{t+m} - v^*\| \leq [\overline{\rho}(t) + \beta c_0 \, \overline{\pi}(t)]/(1-\beta) + 2c_0\beta^m \leq c_2 \cdot \max\{\overline{\rho}(t), \overline{\pi}(t), \beta^m\}. \tag{1}$$

Then making the substitution $s = t + m$ with $t = [s/2]$ and $m = s - t \geq [s/2]$, inequality (1) becomes the inequality in (b):

$$\|\overline{v}_s - v^*\| \leq c_2 \cdot \max\{\overline{\rho}([s/2]), \; \overline{\pi}([s/2]), \; \beta^{[s/2]}\}.$$

To obtain the second part of (b) simply note that if $\rho(t)$ and $\pi(t)$ are nonincreasing, then (1) holds when $\overline{\rho}$ and $\overline{\pi}$ are replaced by ρ and π, respectively. This completes the proof of Theorem 4.8. □

Proof of Theorem 4.9. By Definition 3.8, we have to show that if $\delta = \{g_t\}$ denotes any of the three Markov policies in the statement of the theorem, then

$$\sup_x |\phi(x, g_t(x))| \to 0 \quad \text{as} \quad t \to \infty. \tag{2}$$

Let us consider first the policy $\delta^* = \{f_t^*\}$ defined by the scheme NVI-1. To simplify the notation we shall write $a = f_t^*(x)$; then from 4.4 and Definition 3.2 of ϕ, we have:

$$\begin{aligned}
\phi(x,a) &= \phi(x,a) - v_t^*(x) + v_t^*(x) \\
&= r(x,a) + \beta \int v^*(y) \, q(dy \mid x,a) - v^*(x) \\
&- r_t(x,a) - \beta \int v_t^*(y) \, q_t(dy \mid x,a) + v_t^*(x). \quad [a = f_t^*(x)]
\end{aligned}$$

On the r.h.s., add and subtract $\beta \int v^*(y)\, q_t(dy\,|\,x,a)$, take absolute values, and use the triangle inequality and the definitions of $\rho(t)$ and $\pi(t)$, to obtain:

$$|\phi(x,a)| \le \rho(t) + \beta\|v^*\|\,\pi(t) + \beta\|v_t^* - v^*\| + \|v_t^* - v^*\|. \tag{3}$$

Finally, since $\|v^*\| \le c_0$, use of Theorem 4.8 yields

$$\sup_x |\phi(t, f_t^*(x))| \le [1 + \beta c_0 + (1+\beta)c_1] \cdot \max\{\rho(t), \pi(t)\}, \tag{4}$$

from which (2) follows for $\delta^* = \{f_t^*\}$.

Similarly, for the policy $\bar{\delta} = \{\bar{f}_t\}$, writing

$$\phi(x, \bar{f}_t(x)) = \phi(x, \bar{f}_t(x)) - \bar{v}_t(x) + \bar{v}_t(x),$$

an analogous argument yields

$$|\phi(x, \bar{f}_t(x))| \le \rho(t) + \beta\|v^*\|\,\pi(t) + \beta\|\bar{v}_{t-1} - v^*\| + \|\bar{v}_t - v^*\|,$$

so that (2) follows from Assumption 4.1 and Theorem 4.8(b).

Finally, for $\delta' = \{f_t'\}$, let us write $a = f_t'(x)$ and

$$\phi(x,a) = \phi(x,a) - G_t(x,a,w_t) + G_t(x,a,w_t).$$

On the r.h.s., add and subtract each of the terms

$$T_t w_t(x), \;\; \beta \int v^*(y)\, q_t(dy\,|\,x,a) \;\; \text{and} \;\; v_t^* = T_t v_t^*,$$

to obtain

$$|\phi(x,a)| \le \rho(t) + \beta\|v^*\|\,\pi(t) + \beta\|w_t - v^*\| + \epsilon_t + \beta\|w_t - v_t^*\| + \|v_t^* - v^*\|,$$

with $a = f_t'(x)$, which implies (2) for δ'. □

4.10 Remark. (a) Since the three NVI schemes are defined in terms of the same operators T_t, they are obviously interrelated, and also related to the value-iteration functions $v_t = Tv_{t-1}$ in Remark 2.7; in particular, from that remark and Theorem 4.8, the three sequences $\|v_t^* - v_t\|$, $\|\bar{v}_t - v_t\|$ and $\|w_t - v_t\|$ converge to zero as $t \to \infty$.

It is important to note, on the other hand, the *differences* between the schemes. For instance, to obtain the policy δ^* in NVI-1, at each stage t we have to *solve* equation 4.4, which in general is not a trivial matter. In contrast, the policy δ' in NVI-3 has the advantage that 4.4 does not need to be solved; instead, the functions w_t can be defined in such a way that good *a priori* convergences rates of 4.6 can be obtained, and then $f_t'(x)$ is chosen as any ϵ_t-maximizer of $T_t w_t(x)$.

Finally, with respect to the other two policies, $\bar{\delta}$ in NVI-2 has the advantage that the functions $\bar{v}_t$ in 4.5 are obtained *recursively*, starting "from scratch", and so it seems to be of more direct applicability. It can also be combined with ϵ_t-maximizers, as in NVI-3, which would facilitate computational procedures.

4.10 Remark. (b) We can use equation 3.5 to estimate the "discrepancy" between an optimal policy and each of the policies in Theorem 4.9. For example, take the policy $\delta^* = \{f_t^*\}$. Then equation 3.5 and equation (4) in the proof of Theorem 4.9 yield, for any $n = 0, 1, \ldots$ and $x \in X$,

$$|V_n(\delta^*, x) - E_x^{\delta^*} v^*(x_n)| = \left|\sum_{t=n}^{\infty} \beta^{t-n} E_x^{\delta^*} \phi(x_t, f_t^*(x_t))\right|$$

$$\leq c \sum_{t=n}^{\infty} \beta^{t-n} \max\{\rho(t), \pi(t)\}$$

[where $c := 1 + \beta c_0 + (1+\beta)c_1$]

$$\leq c \cdot \max\{\overline{\rho}(n), \overline{\pi}(n)\}/(1-\beta).$$

This clearly shows that, as was to be expected, the discrepancy between using δ^* and an optimal policy for the MCM $= (X, A, q, r)$ depends on how close to MCM we choose the approximating control models (X, A, q_t, r_t). In Chapter 6 a similar approach is used to estimate the discrepancy, or "deviation from optimality" for several discretization procedures. In the meantime, we illustrate the NVI approach with the following simple application.

Finite-State Approximations

Let (X, A, q, r) be a MCM with *denumerable* state space X. (With the discrete topology, X is a complete, separable and metrizable space, and therefore, a Borel space.) To fix ideas, we shall assume that X is a set of d-dimensional vectors with integer components. We shall write the transition law q as

$$q(y \mid k) := q(\{y\} \mid k), \quad \text{where } k = (x, a) \in \mathbf{K} \text{ and } y \in X,$$

and suppose that Assumptions 2.1 hold. We will present a slight variant of the scheme NVI-2 to obtain finite-state approximations to (X, A, q, r).

Let (X, A, q_t, r_t) be a sequence of approximating control models with $r_t := r$ for all t, and

$$q_t(y \mid k) := q(y \mid k) \cdot 1\{|y| \leq t\} \quad \text{for } t = 1, 2, \ldots,$$

where $1\{\cdot\}$ denotes the indicator function; that is,

$$\begin{aligned} q_t(y \mid k) &:= q(y \mid k) \text{ if } |y| \leq t, \\ &:= 0 \text{ otherwise.} \end{aligned}$$

[Strictly speaking, (X, A, q_t, r_t) might not be a MCM, since possibly $q_t(X \mid k)$ is < 1. But, on the other hand, a look at the proofs of Theorems 4.8 and 4.9 will show that it is not required that q_t be a probability measure; it is only used in fact that $q_t(X \mid k) \leq 1$.] Thus, since $\rho(t) = 0$ for all t, Assumption 4.1 becomes a *tail condition:*

4.11 $$\pi(t) := \sup_{k \in \mathbf{K}} \sum_{|y|>t} q(y \mid k) \to 0 \text{ as } t \to \infty,$$

and equation 4.5 in the scheme NVI-2 results in

$$\overline{v}_t(x) = \max_{a \in A(x)} \left\{ r(x,a) + \beta \sum_{|y| \leq t} \overline{v}_{t-1}(y)\, q(y \mid x, a) \right\}, \quad x \in X.$$

However, since we are looking for *finite*-state approximations of the optimal reward function v^*, we define instead a sequence $\{u_t\}$ by $u_0(\cdot) := 0$, and for $t = 1, 2, \ldots$,

4.12 $$\begin{aligned} u_t(x) &:= \max_{a \in A(x)} \{ r(x,a) + \beta \textstyle\sum_{|y| \leq t} u_{t-1}(y)\, q(y \mid x, a)\} \text{ if } |x| \leq t, \\ &:= 0 \text{ if } |x| > t. \end{aligned}$$

We now define a sequence $\delta'' = \{f_t''\}$ of functions $f_t'' \in \mathbf{F}$ such that $f_0'' \in \mathbf{F}$ is arbitrary, and for $t \geq 1$, $f_t''(x) \in A(x)$ maximizes the r.h.s. of 4.12 if $|x| \leq t$, and $f_t''(x)$ is an arbitrary point in $A(x)$ if $|x| > t$. Then, in this context, the parts corresponding to NVI-2 in Theorems 4.8 and 4.9 yield the following.

4.13 Theorem. *If* 4.11 *holds, then*

(a) $\|u_t - v^*\|_t \leq c_3 \cdot \max\{\pi([t/2]), \beta^{[t/2]}\}$, *where* $\|v\|_t := \sup_{|x| \leq t} |v(x)|$ *and* $c_3 := c_0(2-\beta)/(1-\beta)$.

(b) δ'' *is an ADO policy; in fact,* $\max_{|x| \leq t} |\phi(x, f_t''(x))| \to 0$ *as* $t \to \infty$.

This result is obtained by the same arguments used in the proofs of Theorems 4.8 and 4.9.

Theorem 4.13 is contained in Hernández-Lerma (1986) and it extends a result of D.J. White (1980). Both of these works have been generalized by Cavazos-Cadena (1986, 1987) and applied to the priority assignment problem in queueing systems by Cavazos-Cadena and Hernández-Lerma (1987). Other applications of the scheme NVI-2 are mentioned by Federgruen and Schweitzer (1981, Section 1). Finite-state approximations for *average* cost MCM's are given by Thomas and Stengos (1984), and Acosta Abreu (1987).

2.5 Adaptive Control Models

A MCM $(X, A, q(\theta), r(\theta))$ depending on an ***unknown*** parameter θ is called an ***adaptive*** MCM. [Some authors, e.g., Hinderer (1970), include in this category the case of MCM's with incomplete state information, as in Chapter 4 below.] The main objective of this section is to use the NVI schemes introduced in Section 2.4 to derive ***adaptive*** control policies, i.e., policies combining parameter estimates and control actions.

We begin by re-writing some of the results in Sections 2.2 and 2.3 in terms of the parametric θ-model $(X, A, q(\theta), r(\theta))$. For each (fixed) value of θ, everything remains the same, except for changes in notation:

$$q(\cdot \mid k),\ r(k),\ V(\delta, x),\ V_n(\delta, x),\ v^*(x),\ P_x^\delta,\ E_x^\delta,\ \text{etc.},$$

are changed, respectively, into

$$q(\cdot \mid k, \theta),\ r(k, \theta),\ V(\delta, x, \theta),\ V_n(\delta, x, \theta),\ v^*(x, \theta),\ P_x^{\delta,\theta},\ E_x^{\delta,\theta},\ \text{etc.}$$

We then translate the NVI approximation schemes into appropriate parametric versions (using the "translation rule" 5.4 below) to obtain three adaptive policies whose asymptotic optimality is a consequence of the NVI Theorems 4.8 and 4.9.

Preliminaries

Let Θ be a Borel space, and for each $\theta \in \Theta$, let $(X, A, q(\theta), r(\theta))$ be a MCM satisfying the analogue of Assumptions 2.1, as follows.

5.1 Assumptions.

(a) Same as 2.1(a).

(b) $r(k, \theta)$ is a measurable function on $\mathbf{K}\Theta$ such that $|r(k, \theta)| \le R < \infty$ for all $k = (x, a) \in \mathbf{K}$ and $\theta \in \Theta$, and, moreover, $r(x, a, \theta)$ is a continuous function of $a \in A(x)$ for every $x \in X$ and $\theta \in \Theta$.

(c) $q(\cdot \mid k, \theta)$ is a stochastic kernel on X given $\mathbf{K}\Theta$ such that

$$\int v(y, \theta)\, q(dy \mid x, a, \theta)$$

is a continuous function of $a \in A(x)$ for every $x \in X$, $\theta \in \Theta$, and $v \in B(X\Theta)$.

Under these assumptions, which are supposed to hold throughout the following, all the results in Sections 2.2 and 2.3 hold for each $\theta \in \Theta$. In particular, introducing the reward functions

$$V(\delta, x, \theta) := E_x^{\delta,\theta} \sum_{t=0}^{\infty} \beta^t r(x_t, a_t, \theta)$$

and

$$v^*(x,\theta) := \sup_{\delta\in\Delta} V(\delta,x,\theta),$$

we can re-state Theorem 2.2 and Proposition 3.4 in combined form as follows.

5.2 Theorem. (a) *v^* is the unique solution in $B(X\Theta)$ of the (θ-) DPE*

$$v^*(x,\theta) = \max_{a\in A(x)} \left\{ r(x,a,\theta) + \beta \int_X v^*(y,\theta)\, q(dy\,|\,x,a,\theta) \right\},$$

or equivalently,

$$\max_{a\in A(x)} \phi(x,a,\theta) = 0 \quad \textit{for all } x\in X \textit{ and } \theta\in\Theta,$$

where ϕ is the function on $\mathbf{K}\Theta$ defined by

$$\phi(x,a,\theta) := r(x,a,\theta) + \beta \int v^*(y,\theta)\, q(dy\,|\,x,a,\theta) - v^*(x,\theta).$$

(b) *For each $\theta\in\Theta$, a stationary policy $f^*(\cdot,\theta)\in\mathbf{F}$ is (θ-) discount optimal, i.e., $V[f^*(\cdot,\theta),x,\theta] = v^*(x,\theta)$ for every $x\in X$, if and only if $\phi(x,f^*(x,\theta),\theta) = 0$ for every $x\in X$.*

Remark. Since X, A, $\mathbf{K}$ and Θ are Borel spaces, the products $XA\Theta$ and $\mathbf{K}\Theta$ are also Borel spaces [Bertsekas and Shreve (1978), p. 119]. Thus, as in Remark 2.3, the existence [and measurability in (x,θ)] of a stationary policy $f^*(x,\theta)$ as in 5.2(b) is insured under Assumptions 5.1.

All the other results in Sections 2.2 and 2.3 are also valid (for each θ) with the obvious changes in notation. For instance, Theorem 3.6 becomes: a policy δ is θ-ADO (i.e., ADO for the θ-MCM) if and only if

5.3 $\quad \phi(x_t,a_t,\theta) \to 0$ in probability-$P_x^{\delta,\theta}$ for every $x\in X$,

where, as in Definition 1.3, θ-ADO means that, for every $x\in X$,

$$|V_n(\delta,x,\theta) - E_x^{\delta,\theta} v^*(x_n,\theta)| \to 0 \quad \text{as } n\to\infty.$$

Nonstationary Value-Iteration

If $\theta\in\Theta$ is the "true" (but unknown) parameter value we can approximate the optimal reward function $v^*(x,\theta)$ and obtain θ-ADO policies via appropriate versions of the NVI schemes in Section 2.4. The idea is to consider sequences

5.4 $\quad r_t(k) := r(k,\theta_t)$ and $q_t(\cdot\,|\,k) := q(\cdot\,|\,k,\theta_t), \quad t = 0,1,\ldots,$

where $k = (x,a)\in\mathbf{K}$ and $\{\theta_t\}$ is a sequence in Θ converging to θ. Thus, Assumption 4.1 is obviously replaced by the following.

5.5 Assumption. For any $\theta \in \Theta$ and any sequence $\{\theta_t\}$ in Θ such that $\theta_t \to \theta$, both $\rho(t,\theta)$ and $\pi(t,\theta)$ converge to zero as $t \to \infty$, where

$$\rho(t,\theta) := \sup_k |r(k,\theta_t) - r(k,\theta)|$$

and

$$\pi(t,\theta) := \sup_k \|q(\cdot \mid k,\theta_t) - q(\cdot \mid k,\theta)\|.$$

As in Remark 4.2, we define the non-increasing sequences

$$\overline{\rho}(t,\theta) := \sup_{s \geq t} \rho(s,\theta) \quad \text{and} \quad \overline{\pi}(t,\theta) := \sup_{s \geq t} \pi(s,\theta).$$

Assumption 5.5 is a condition of continuity in the parameter $\theta \in \Theta$ uniform in $k \in \mathbf{K}$, and one would expect that it implies continuity of $v^*(x,\theta)$ in θ. This is indeed the case and the continuity is uniform on X; this is obtained from the following result.

5.6 Proposition. *For any θ and θ_t in Θ,*

$$\|v^*(\cdot,\theta_t) - v^*(\cdot,\theta)\| \leq c_1 \cdot \max\{\rho(t,\theta), \pi(t,\theta)\},$$

where c_1 is the constant in 4.7.

Actually this proposition is *exactly the same* as Theorem 4.8(a) under the substitution 5.4, when

$$v_t^*(x),\; v^*(x),\; \rho(t) \;\text{ and }\; \pi(t)$$

in that theorem are replaced, respectively, by

$$v^*(x,\theta_t),\; v^*(x,\theta),\; \rho(t,\theta) \;\text{ and }\; \pi(t,\theta).$$

Thus, Proposition 5.6 is an illustration of how one translates the approximation results in Section 2.4 to the "adaptive" results in the present section.

To complete the "translation", let us write the θ-DPE in Theorem 5.2 in terms of the DP operator T_θ on $B(X)$ defined by

5.7 $$T_\theta v(x) := \max_{a \in A(x)} G(x,a,\theta,v),$$

for every function v in either $B(X)$ or $v(x,\theta)$ in $B(X\Theta)$, where

$$G(k,\theta,v) := r(k,\theta) + \beta \int v(y)\, q(dy \mid k,\theta) \quad \text{for } k = (x,a) \in \mathbf{K}.$$

For each $\theta \in \Theta$, the fixed point $v^*(\cdot,\theta) \in B(X)$ of T_θ is the optimal reward function for the θ-MCM $(X, A, q(\theta), r(\theta))$, and we write the θ-DPE as

$$v^*(x,\theta) = T_\theta v^*(x,\theta) \quad \text{for } x \in X.$$

Finally, given a sequence $\{\theta_t\}$ in Θ we define the operators

$$T_t := T_{\theta_t}, \quad \text{with} \quad G_t(k,v) := G(k,\theta_t,v),$$

i.e.,

$$\textbf{5.8} \quad \begin{aligned} T_t v(x) &:= \max_{a\in A(x)} G_t(x,a,v) \\ &= \max_{a\in A(x)}\{r(x,a,\theta_t) + \beta \int v(y)\, q(dy \mid x,a,\theta_t)\}. \end{aligned}$$

With this notation, the NVI schemes in Section 2.4 become as follows.

NVI-1. For each $t = 0, 1, \ldots$, let $v_t^*(\cdot) \equiv v^*(\cdot,\theta_t) \in B(X)$ be the fixed point of T_t, and let $\delta_\theta^* = \{f_t^*\}$ be a sequence of functions $f_t^*(\cdot) \equiv f^*(\cdot,\theta_t)$ in **F** which maximize the r.h.s. of 5.8, i.e.,

$$v^*(x,\theta_t) = G_t[x, f^*(x,\theta_t), v^*(\cdot,\theta_t)].$$

NVI-2. For each $t = 0, 1, \ldots$, let $\overline{v}_t(\cdot) \equiv \overline{v}_t(\cdot,\theta_t)$ be functions in $B(X)$ defined recursively by $\overline{v}_t := T_t\overline{v}_{t-1}$, i.e., for every $x \in X$ and $t \geq 0$,

$$\textbf{5.9} \quad \begin{aligned} \overline{v}_t(x,\theta_t) &:= T_t\overline{v}_{t-1}(x,\theta_{t-1}) \\ &= \max_{a\in A(x)}\{r(x,a,\theta_t) + \beta \int \overline{v}_{t-1}(y,\theta_{t-1}) q(dy \mid x,a,\theta_t)\}, \end{aligned}$$

with $\overline{v}_{-1}(\cdot) := 0$. Let $\overline{\delta}_\theta = \{\overline{f}_t\}$ be a sequence of functions $\overline{f}_t(\cdot) \equiv \overline{f}_t(\cdot,\theta_t)$ in **F** such that $\overline{f}_t(x,\theta_t)$ maximizes the r.h.s. of 5.9 for every $x \in X$ and $t \geq 0$. [Notice that both $\overline{v}_t$ and $\overline{f}_t$ depend, not only on θ_t but on all the values $\theta_0, \theta_1, \ldots, \theta_t$; however, we shall keep the shorter notation $\overline{v}_t(x,\theta_t)$ and $\overline{f}_t(x,\theta_t)$.]

NVI-3. Let $w_t(\cdot) \equiv w_t(\cdot,\theta_t)$ be functions in $B(X)$ such that

$$\|w_t(\cdot,\theta_t) - v^*(\cdot,\theta_t)\| \to 0 \quad \text{as} \quad t \to \infty,$$

and let $\{\epsilon_t\}$ be a decreasing sequence of positive numbers converging to zero. For each $x \in X$ and $t \geq 0$, let

$$A_t(x) := \{a \in A(x) \mid G(x,a,\theta_t,w_t) \geq T_t w_t(x) - \epsilon_t\},$$

and let $\delta' = \{f_t'\}$ be a sequence of measurable functions $f_t'(\cdot) \equiv f_t'(\cdot,\theta_t)$ from X to A such that $f_t'(x) \in A_t(x)$ for every $x \in X$ and $t \geq 0$.

As a consequence of Theorems 4.8 and 4.9, we then obtain the following.

5.10 Corollary. *Suppose that Assumptions* 5.1 *and* 5.5 *hold, and let* $\theta_t \to \theta$. *Then each of the sequences* $v^*(x,\theta_t)$, $\overline{v}_t(x,\theta_t)$ *and* $w_t(x,\theta_t)$ *converges to* $v^*(x,\theta)$ *uniformly in* $x \in X$; *the inequalities in Theorem* 4.8 *also hold in the present case, with* $\rho(t) = \rho(t,\theta)$ *and* $\pi(t) = \pi(t,\theta)$ *as in* 5.5. [*Compare, for instance, Theorem* 4.8(a) *and Proposition* 5.6.] *Moreover, each of the policies* δ_θ^*, $\overline{\delta}_\theta$ *and* δ_θ' *is uniformly* θ*-ADO, that is, as* $t \to \infty$, *each of the sequences*

$$\sup_x |\phi(x, f^*(x,\theta_t), \theta)|,$$

$$\sup_x |\phi(x, \overline{f}_t(x, \theta_t), \theta)|,$$

and

$$\sup_x |\phi(x, f_t'(x, \theta_t), \theta)|$$

converges to zero.

The Principle of Estimation and Control

All the remarks in 4.10 concerning the NVI policies are valid, of course, in the present parametric case for δ_θ^*, $\overline{\delta}_\theta$ and δ_θ'. In addition, we can now relate the scheme NVI-1 with the so-called "principle of estimation and control (PEC)". The latter is another name for what Mandl (1974) called the "method of substituting the estimates into optimal stationary controls", and which, except perhaps for small variations, is also found in the literature on stochastic adaptive control under the names of "Naive feedback controller", "Certainty-equivalence controller", "Self-tuning regulator", or "Self-optimizing controls". A PEC (adaptive) policy is formally defined below (in 5.15), but the idea to construct it is the following.

5.11 Construction of the PEC Policy.

(a) For each "admissible" value of the parameter $\theta \in \Theta$, "solve" the θ-DPE (Theorem 5.2); i.e., find an optimal stationary policy $f^*(\cdot, \theta) \in \mathbf{F}$ such that $f^*(x, \theta)$ maximizes the r.h.s. of the θ-DPE for every $x \in X$.

(b) At each time t, compute an estimate $\theta_t \in \Theta$ of θ^*, where θ^* is assumed to be the true—but unknown—parameter value. Thus the "true" optimal stationary policy if $f^*(\cdot, \theta^*) \in \mathbf{F}$, and the optimal action at time t is $a_t = f^*(x_t, \theta^*)$. However, we do not know the true parameter value, and therefore, we choose instead, at time t, the control

$$a_t^* := f^*(x_t, \theta_t);$$

in other words, we simply "substitute the estimates into optimal stationary controls".

What we want to emphasize here is that the PEC policy $\{a_t^*\}$ thus constructed is the "same" as the NVI-1 policy δ_θ^*, with $\theta = \theta^*$, and furthermore, the asymptotic optimality of δ_θ^* when $\theta_t \to \theta$ is a direct consequence of the *continuity* Assumption 5.5. More precisely, we have that Assumption 5.5 implies Proposition 5.6, and the two combined imply that δ_θ^* is θ-ADO, since "exactly" as in the proof of Theorem 4.9 [equation (3)] it can be shown that

5.12 $|\phi(x, f^*(x, \theta_t), \theta)| \leq \rho(t, \theta) + \beta c_0 \pi(t, \theta) + (1 + \beta) \| v^*(\cdot, \theta_t) - v^*(\cdot, \theta) \|$

for every $x \in X$.

5.13 Remark (Lipschitz-Continuity of $v^*(x,\theta)$ in θ). Instead of the continuity Assumption 5.5 (all other assumptions remaining the same), let us suppose that r and q satisfy Lipschitz conditions in θ: There are constants L_1 and L_2 such that for every $k = (x,a)$ in $\mathbf{K}$ and every θ and θ' in Θ,

$$|r(k,\theta) - r(k,\theta')| \leq L_1 d(\theta,\theta'),$$

and

$$\|q(\cdot \mid k,\theta) - q(\cdot \mid k,\theta')\| \leq L_2 d(\theta,\theta'),$$

where d is the metric on Θ. Then

$$\|v^*(\cdot,\theta) - v^*(\cdot,\theta')\| \leq L_3 d(\theta,\theta'),$$

where $L_3 := (1-\beta)^{-1}(L_1 + \beta c_0 L_2)$. The proof is similar to that of Proposition 5.6 [or Theorem 4.8(a)]; cf. Kolonko (1983), Hernández-Lerma and Cavazos-Cadena (1988), or Rieder and Wagner (1986). [In Theorem 6.7 below we show that $v^*(x,\theta)$ is Lipschitz in x, rather than θ.]

Adaptive Policies

We have essentially all the ingredients to define adaptive policies, i.e., policies combining parameter estimates and control actions, except for one thing: we have not said yet how to estimate parameters. There are several ways in which one can estimate parameters in controlled stochastic systems; some of these techniques will be presented in Chapter 5. For the moment, however, for the sake of continuity in the exposition, we will simply assume that the estimators are given and that they are sufficiently "robust", in the following sense.

5.14 Definition. A sequence $\{\hat{\theta}_t\}$ of measurable functions $\hat{\theta}_t$ from H_t to Θ is said to be a sequence of *strongly consistent* (SC) *estimators* of θ if, as $t \to \infty$, $\hat{\theta}_t = \hat{\theta}_t(h_t)$ converges to θ $P_x^{\delta,\theta}$-almost surely (a.s.) for every $x \in X$ and $\delta \in \Delta$.

Now, for $\theta \in \Theta$, let $\delta_\theta^* = \{f_t^*\}$, $\overline{\delta}_\theta = \{\overline{f}_t\}$ and $\delta' = \{f_t'\}$ be the policies defined by the schemes NVI-1,2 and 3, respectively, and let $\{\hat{\theta}_t\}$ be a sequence of measurable functions $\hat{\theta}_t$ from H_t to Θ. We use this notation to define the following adaptive policies.

5.15 Definition.

(a) Let $\delta^* = \{\delta_t^*\}$ be the policy defined by

$$\delta_t^*(h_t) := f^*(x_t, \hat{\theta}_t(h_t)) \quad \text{for} \quad h_t \in H_t \quad \text{and} \quad t \geq 0.$$

We call δ^* a *PEC adaptive policy* (cf. 5.11).

(b) The policy $\overline{\delta} = \{\overline{\delta}_t\}$ defined by

$$\overline{\delta}_t(h_t) := \overline{f}_t(x_t, \hat{\theta}_t(h_t)) \quad \text{for} \quad h_t \in H_t \quad \text{and} \quad t \geq 0,$$

is called an *NVI adaptive policy*.

(c) The policy $\delta' = \{\delta'_t\}$ defined by

$$\delta'_t(h_t) := f'_t(x_t, \hat{\theta}_t(h_t)) \quad \text{for} \quad h_t \in H_t \quad \text{and} \quad t \geq 0,$$

is called a *modified-PEC adaptive policy.*

Suppose now we want to prove that (e.g.) the PEC adaptive policy δ^* is θ-ADO; then, by 5.3, it suffices to verify that, as $t \to \infty$,

5.16 $$|\phi(x_t, f^*(x_t, \hat{\theta}_t), \theta)| \to 0 \quad P_x^{\delta,\theta}\text{-a.s.} \quad \text{for every} \quad x \in X,$$

where we have written $\hat{\theta}_t(h_t)$ as $\hat{\theta}_t$. But on the other hand, we already know (Corollary 5.10) that, under Assumption 5.5,

$$\sup_{x \in X} |\phi(x, f^*(x, \theta_t), \theta)| \to 0$$

for *any* sequence θ_t that converges to θ. Thus the latter implies 5.16 *if* $\{\hat{\theta}_t\}$ is a sequence of SC estimators of θ. The *same* argument applies to verify the θ-asymptotic discount optimality of the NVI and the modified-PEC adaptive policies, and therefore, we conclude the following.

5.17 Corollary. *Suppose that the assumptions of Corollary* 5.10 *hold and that* $\{\hat{\theta}_t\}$ *is a sequence of SC estimators of* θ. *Then each of the adaptive policies* δ^*, $\bar{\delta}$ *and* δ' *in* 5.15 *are* θ*-ADO.*

We have thus shown how to derive ADO adaptive policies using the NVI approximation schemes in Section 2.4, which is a recurrent theme in these notes. The same approach can be used to study adaptive control problems for other types of stochastic processes, e.g., semi-Markov, partially observable, etc., some of which are studied in later chapters. In particular, we will consider in the following section the important case of MCM's with i.i.d. disturbances with unknown distribution.

2.6 Nonparametric Adaptive Control

We now consider in this section a MCM (X, A, q, r) whose transition law q is defined by a discrete-time dynamic model

6.1 $$x_{t+1} = F(x_t, a_t, \xi_t) \quad \text{for} \quad t = 0, 1, \ldots; \quad x_0 \text{ given.}$$

Here the *disturbance* (or driving) process $\{\xi_t\}$ is a sequence of i.i.d. random elements (independent of x_0) with values in a Borel space S, and *unknown* distribution $\theta \in \mathbf{P}(S)$, where $\mathbf{P}(S)$ is the space of probability measures on S. (Since S is a Borel space, $\mathbf{P}(S)$ is also a Borel space; see Appendix B.) Thus the unknown "parameter" is the disturbance distribution θ, and the transition law

$$q(B \mid k, \theta) = \text{Prob}(x_{t+1} \in B \mid x_t = x, a_t = a), \quad \text{where} \quad k = (x, a) \in \mathbf{K},$$

is given, for $B \in \mathcal{B}(S)$, by

$$q(B \mid k, \theta) = \int_S 1_B[F(k, s)]\, \theta(ds) = \theta(\{s \in S \mid F(k, s) \in B\}). \tag{6.2}$$

The function $F : \mathbf{K}S \to X$ in 6.1 is assumed to be measurable, of course.

In some cases—see, e.g., the inventory/production system in Section 1.3—it is convenient to allow the reward function r to depend on the "disturbances", and therefore, we let r to be of the form

$$r(k, \theta) = \int_S \overline{r}(k, s)\, \theta(ds) \quad \text{for} \quad k \in \mathbf{K}, \tag{6.3}$$

where $\overline{r} \in B(\mathbf{K}S)$; that is, r is the expected value

$$r(x, a, \theta) = E^{\theta}[\overline{r}(x_t, a_t, \xi_t) \mid x_t = x, a_t = a].$$

In this section, we consider the MCM $(X, A, q(\theta), r(\theta))$ and show first that the setting of Section 2.5 is *not* the appropriate one, in general. We then show how things can be changed in order to obtain θ-ADO adaptive policies.

The Parametric Approach

Let Θ be the set of "admissible" disturbance distributions, a Borel subset of $\mathbf{P}(S)$, and consider the adaptive MCM $(X, A, q(\theta), r(\theta))$, with $q(\theta)$ and $r(\theta)$ given by 6.2 and 6.3, respectively, for $\theta \in \Theta$. Let us suppose for a moment that conditions on $F(x, a, s)$ and $\overline{r}(x, a, s)$ are imposed so that Assumptions 5.1 hold. In such a case, the results in Section 2.5 (e.g., Corollary 5.10) hold *if* Assumption 5.5 is satisfied. To obtain the latter, we see that, from 6.3 and inequality B.1 in Appendix B,

$$\begin{aligned} |r(k, \theta_t) - r(k, \theta)| &= \left| \int \overline{r}(k, s)\, \{\theta_t(ds) - \theta(ds)\} \right| \\ &\le R\|\theta_t - \theta\| \quad \text{for all} \quad k \in \mathbf{K}, \end{aligned}$$

where $\|\theta_t - \theta\|$ is the variation norm of the finite signed measure $\theta_t - \theta$. Thus $\rho(t, \theta)$ in 5.5 satisfies

$$\rho(t, \theta) \le R\|\theta_t - \theta\|.$$

Similarly, from 6.2 and B.2 in Appendix B, we obtain

$$\|q(\cdot \mid k, \theta_t) - q(\cdot \mid k, \theta)\| \le \|\theta_t - \theta\| \quad \text{for all} \quad k \in \mathbf{K},$$

so that

$$\pi(t, \theta) \le \|\theta_t - \theta\|.$$

Thus Assumption 5.5 holds if the probability distributions θ_t are "estimates" of θ that satisfy

6.4 $$\|\theta_t - \theta\| \to 0 \text{ a.s. as } t \to \infty.$$

And here is precisely where the difficulty lies with the "parametric" approach of Section 2.5; namely, 6.4 is very strong requirement. That is, non-parametric statistical estimation methods indeed yield "consistent" estimates, but typically in forms weaker than in variation norm. There are special cases, of course, in which 6.4 holds. For instance, this is shown to be the case in a situation briefly described in Section 2.7, where the disturbance set S is $\mathbf{R}^d$ and the distribution θ is absolutely continuous; but then again, this situation excludes many important applications, e.g., when θ is discrete—as in queueing systems.

On the other hand, in the general case, with Borel disturbance space S and arbitrary distribution θ, the best approach seems to be to use the empirical distribution to estimate θ, but then we do not get 6.4—except in special cases, e.g., when θ is discrete. Therefore, to study the general case, we avoid 6.4 with a slight modification of the "parametric" approach. To do this, we start by imposing a different set of assumptions on the control system 6.1.

New Setting

Let d_1, d_2 and d_3 denote, respectively, the metrics on X, A and S, and let d be the metric on $\mathbf{K}$ defined by $d := \max\{d_1, d_2\}$. We suppose the following.

6.5 Assumptions. There are constants R, L_0, L_1 and L_2 such that

(a) $|\bar{r}(k,s)| \leq R$ and $|\bar{r}(k,s) - \bar{r}(k',s)| \leq L_0 d(k,k')$ for every k and k' in $\mathbf{K}$ and all $s \in S$.

(b) $A(x)$ is a compact subset of A for every $x \in X$, and

$$H(A(x), A(x')) \leq L_1 d_1(x,x') \quad \text{for every } x \text{ and } x' \text{ in } X,$$

where H is the Hausdorff metric (Appendix D).

(c) $\|q(\cdot \mid k,\theta) - q(\cdot \mid k',\theta)\| \leq L_2 d(k,k')$ for every k and k' in $\mathbf{K}$ and $\theta \in \Theta$.

(d) The function $F(k,s)$ in 6.1 is continuous in $k \in \mathbf{K}$ and, moreover, the family of functions $\{F(k,\cdot), k \in \mathbf{K}\}$ is equicontinuous at each point s in S; that is, for each $s \in S$ and $\epsilon > 0$, there exists $\gamma > 0$ such that

$$d_3(s,s') < \gamma \text{ implies } d_1[F(k,s), F(k,s')] < \epsilon \text{ for all } k \in \mathbf{K}.$$

Comments on these assumptions are given at the end of this section. Right now, what it needs to be remarked is that they are introduced because, in the new approach, we need the optimal reward function $v^*(x,\theta) \equiv v^*(x)$ to be Lipschitz-continuous in x. Therefore, to begin with, we consider the DP operator T_θ in 5.7 to be defined, not on $B(X)$, but on the space $C(X)$ of bounded *continuous* functions on X:

6.6 $$T_\theta v(x) := \max_{a \in A(x)} G(x,a,\theta,v), \quad \text{with } v \in C(X),$$

where

$$\begin{aligned} G(k,\theta,v) &:= r(k,\theta) + \beta \int_X v(y)\, q(dy \mid k,\theta) \\ &= r(k,\theta) + \beta \int_S v[F(k,s)]\, \theta(ds). \end{aligned}$$

[The latter equality results from the "change of variable" integration formula; see, e.g., Ash (1972, p. 225), or Royden (1968, p. 318).] We can now state the following.

6.7 Theorem. *Suppose that Assumptions 6.5 hold and let us write $v^*(x,\theta)$ as $v^*_\theta(x)$. Then, for each $\theta \in \Theta$.*

(a) *v^*_θ is the unique solution in $C(X)$ of the DPE*

$$v^*_\theta(x) = T_\theta v^*_\theta(x) \quad \textit{for every } x \in X.$$

(b) *$|v^*_\theta(x) - v^*_\theta(x')| \le L^* \cdot d_1(x,x')$ for every x and x' in X, where*

$$L^* = (L_0 + \beta L_2 c_0) \cdot \max\{1, L_1\},$$

and c_0 is the constant in 4.7, an upper bound for $\|v^*_\theta\|$.

(c) *The family of functions*

$$V^* := \{v^*_\theta[F(k,\cdot)],\ k \in \mathbf{K}\}$$

is uniformly bounded and equicontinuous at each point s in S.

Proof. Part (a) is obtained from Theorem 2.8. Part (b) is a special case of Lemma 6.8 below, taking $j(x) = v^*_\theta(x)$ and $J(k) = G(k,\theta,v^*_\theta)$, and noting that from 6.5(a) and (c), such a function $J(k)$ is Lipschitz with constant $L = (L_0 + \beta L_2 c_0)$. Part (c) follows from (b) and Assumption 6.5(d). □

6.8 Lemma. *Suppose that Assumptions 6.5 hold, and let $J(k) = J(x,a)$ be a real-valued Lipschitz function on $\mathbf{K}$. Then*

$$j(x) := \max_{a \in A(x)} J(x,a)$$

is Lipschitz on X. More precisely, if L is a constant satisfying

$$|J(k) - J(k')| \le L \cdot d(k,k') \quad \textit{for every } k \textit{ and } k' \textit{ in } \mathbf{K},$$

then

$$|j(x) - j(x')| \le L^* \cdot d_1(x,x'), \quad \textit{where} \quad L^* := L \cdot \max\{1, L_1\}.$$

Proof of the Lemma. Let f be a measurable function from X to A such that $f(x) \in A(x)$ and $j(x) = J(x, f(x))$ for every x in X (see Appendix D, Proposition D.3), and using 6.5(b), choose a' in $A(x')$ such that

$$d_2(f(x), a') \leq L_1 \cdot d_1(x, x').$$

Therefore, since $j(x') \geq J(x', a)$ for every $a \in A(x')$,

$$\begin{aligned} j(x) - j(x') &\leq J(x, f(x)) - J(x', a') \\ &\leq L \cdot d[(x, f(x)), (x', a')] \\ &= L \cdot \max\{d_1(x, x'), d_2(f(x), a')\} \\ &\leq L^* \cdot d_1(x, x'). \end{aligned}$$

Similarly, we obtain $j(x') - j(x) \leq L^* \cdot d_1(x, x')$. □

The need to consider such a strong result as 6.7(c) will be clear below, when we consider the problem of estimating the unknown disturbance distribution θ.

The Empirical Distribution Process

To estimate θ we will use the empirical distribution $\{\theta_t\}$ of the disturbance process $\{\xi_t\}$, defined by

$$\textbf{6.9} \quad \theta_t(B) := t^{-1} \sum_{i=0}^{t-1} 1_B(\xi_i), \quad \text{for all } t \geq 1 \text{ and } B \in \mathcal{B}(S).$$

We assume that $\theta_t \in \Theta$ for all t. Now, for each Borel set B in S, the random variables $1_B(\xi_i)$ are i.i.d. with mean $E\,1_B(\xi_i) = \theta(B)$, and therefore, by the Strong Law of Large Numbers

$$\theta_t(B) \to \theta(B) \text{ a.s. as } t \to \infty.$$

Moreover, if θ is a *discrete* distribution, Scheffe's Theorem (see Appendix B) implies that 6.4 holds for the empirical process θ_t in 6.9, but this is not true for general θ. What we do know [Gaenssler and Stute (1979), p. 211] is that θ_t *converges weakly* to θ a.s., that is, at $t \to \infty$,

$$\int h\, d\theta_t \to \int h\, d\theta \text{ a.s. for every } h \in C(S);$$

in particular, since $v^*_\theta[F(k, s)]$ is continuous in $s \in S$, then

$$\textbf{6.10} \quad \int_S v^*_\theta[F(k, s)]\theta_t(ds) \to \int_S v^*_\theta[F(k, s)]\theta(ds) \text{ a.s. for } \textit{each } k \in \mathbf{K}.$$

But this still is not good enough for the adaptive control results we want. What we actually need is the convergence in 6.10 to hold *uniformly* in $k \in \mathbf{K}$, and here is where Theorem 6.7(c) comes in: Since V^* is uniformly bounded and equicontinuous at each point s in S, it follows from Proposition B.8 in Appendix B that V^* is a θ-uniformity class (Definition B.7) for every $\theta \in \Theta$, and therefore, 6.10 holds uniformly on $\mathbf{K}$, that is,

6.11 $$\eta(t,\theta) \to 0 \text{ a.s. as } t \to \infty,$$

where

6.12 $$\begin{aligned}\eta(t,\theta) &:= \sup_{k\in\mathbf{K}} \left| \int v_\theta^*(y)\, q(dy \mid k, \theta_t) - \int v_\theta^*(y)\, q(dy \mid k, \theta) \right| \\ &= \sup_k \left| \int v_\theta^*[F(k,s)]\, \theta_t(ds) - \int v_\theta^*[F(k,s)]\, \theta(ds) \right| \\ &= \sup_k \left| t^{-1} \sum_{i=0}^{t-1} v_\theta^*[F(k,\xi_i)] - \int v_\theta^*[F(k,s)]\, \theta(ds) \right| .\end{aligned}$$

Similarly, if in addition to 6.5(a) we assume that the family of functions

$$\mathcal{R} := \{\bar{r}(k,\cdot), k \in S\} \text{ is equicontinuous at each } s \in S,$$

then $\mathcal{R}$ is a θ-uniformity class for every $\theta \in \Theta$, and therefore (Proposition B.8),

6.13 $$\rho(t,\theta) \to 0 \text{ a.s. as } t \to \infty,$$

where (cf. 5.5)

$$\begin{aligned}\rho(t,\theta) &:= \sup_{k\in\mathbf{K}} |r(k,\theta_t) - r(k,\theta)| \\ &= \sup_k \left| \int \bar{r}(k,s)\, \theta_t(ds) - \int \bar{r}(k,s)\, \theta(ds) \right| .\end{aligned}$$

We can summarize these results as follows.

6.14 Lemma. *Under Assumption* 6.5,

(a) 6.11 *holds, and*

(b) *if in addition* $\mathcal{R}$ *is equicontinuous at each point* $s \in S$, *then* 6.13 *also holds.*

Nonparametric Adaptive Policies

Conditions 6.11 and 6.13 are the "nonparametric" analogue of Assumption 5.5. Thus, if we change Assumptions 5.1 and 5.5 by the assumptions of Lemma 6.14, then the conclusions of Corollaries 5.10 and 5.17 hold in the present nonparametric case when $\pi(t,\theta)$ is replaced by $\eta(t,\theta)$ in 6.12 and the constants c_1 and c_2 in 4.7 are replaced, respectively, by

$$c_1' := (1+\beta)/(1-\beta) \text{ and } c_2' := c_1' + 2c_0.$$

The precise result is obtained as Corollaries 5.10 and 5.17, i.e., from Theorems 4.8 and 4.9, and it can be stated as follows.

6.15 Theorem. *Suppose that the assumptions of Lemma* 6.14(a) *and* (b) *hold, and let* $v^*(\cdot,\theta_t)$, $\overline{v}_t(\cdot,\theta_t)$ *and* $w_t(\cdot,\theta_t)$ *be the functions in* NVI-1,2,3 (*Section* 5), *when the DP operator* T_θ *is defined on* $C(X)$ (cf. 6.6), *and* $\{\theta_t\}$ *is the empirical distribution process in* 6.9. *Then*

(a) $\|v^*(\cdot,\theta_t) - v^*(\cdot,\theta)\| \le c_1' \cdot \max\{\rho(t,\theta), \eta(t,\theta)\}$.

(b) $\|\overline{v}_t(\cdot,\theta_t) - v^*(\cdot,\theta)\| \le c_1' \cdot \max\{\overline{\rho}([t/2],\theta),\ \overline{\eta}([t/2],\theta),\ \beta^{[t/2]}\}$, where

$$\overline{\rho}(t,\theta) := \sup_{i\ge t} \rho(i,\theta) \quad\text{and}\quad \overline{\eta}(t,\theta) := \sup_{i\ge t} \eta(i,\theta).$$

[Cf. Theorem 4.8(b).]

(c) $\|w_t(\cdot,\theta_t) - v^*(\cdot,\theta)\| \le \|w_t(\cdot,\theta_t) - v^*(\cdot,\theta_t)\| + \|v^*(\cdot,\theta_t) - v^*(\cdot,\theta)\|$.

Moreover, the adaptive policies δ^*, $\overline{\delta}$ *and* δ' *in* 5.15 (*with the above changes*) *are* θ-*ADO*.

Remarks on the Assumptions 6.5. Assumption 6.5(b) holds, e.g., if $A(x) = A$ is compact and independent of $x \in X$. It also holds in the inventory/production system of Section 1.3, in which $A(x)$ is the interval $[0, C - x]$. In such a case, the definition of the Hausdorff metric (Appendix D) yields

$$H(A(x), A(x')) = |x - x'| \quad\text{for every } x \text{ and } x' \text{ in } X = [0, C].$$

Assumptions 6.5(a) and (d) are also verified in the inventory/production example. More generally, 6.5(d) trivially holds in the additive-noise case, say,

$$F(x,a,s) = b(x,a) + s \quad\text{or}\quad b(x,a) + c(x)s$$

if b and c are continuous functions and c is bounded. With respect to 6.5(c) we have the following.

6.16 Proposition. *Each of the following conditions implies* 6.5(c).

(a) *There is a constant* L *such that, for every* k *and* k' *in* $\mathbf{K}$ *and* $\theta \in \Theta$,

$$\sup_{B\in\mathcal{B}(X)} \theta(B[k]\,\Delta B[k']) \le L \cdot d(k,k'),$$

where $B[k] := \{s \in S \mid F(k,s) \in B\}$, *and* Δ *denotes the symmetric difference of sets.*

(b) *For every* $k \in \mathbf{K}$ *and* $\theta \in \Theta$, $q(\cdot \mid k,\theta)$ *has a density* $p_\theta(\cdot \mid k)$ *with respect to a sigma-finite measure* μ *on* X *such that, for every* $x \in X$, $\theta \in \Theta$, *and* k *and* k' *in* $\mathbf{K}$,

$$|p_\theta(x \mid k) - p_\theta(x \mid k')| \le L(x) \cdot d(k,k'),$$

where $L(x)$ *is a* μ-*integrable function.*

Proof. (a) From 6.2 and equality B.2 in Appendix B,

$$\begin{aligned}
\|q(\cdot \mid k, \theta) - q(\cdot \mid k', \theta)\| &= 2 \sup_B |q(B \mid k, \theta) - q(B \mid k', \theta)| \\
&= 2 \sup_B |\theta(B[k]) - \theta(B[k'])| \\
&\leq 2 \sup_B \theta(B[k] \Delta B[k']) \\
&\leq 2L\, d(k, k').
\end{aligned}$$

(b) This part follows from B.3 in Appendix B, according to which

$$\begin{aligned}
\|q(\cdot \mid k, \theta) - q(\cdot \mid k', \theta)\| &= \int |p_\theta(x \mid k) - p_\theta(x \mid k')| \mu(dx) \\
&\leq \left(\int L\, d\mu \right) \cdot d(k, k'). \qquad \square
\end{aligned}$$

In most of the standard (finite-dimensional) control systems, the disturbance distribution θ is either absolutely continuous with respect to Lebesgue measure or discrete—e.g., Gaussian, gamma, Poisson, etc.—and one can try to verify 6.5(c) using Proposition 6.16(b). For instance, if X and A are finite-dimensional Euclidean spaces and $p_\theta(\cdot \mid k)$ is continuously differentiable in k with bounded derivative, then 6.16(b) is obtained from the Mean Value Theorem if $L(x)$ is μ-integrable; here $L(x)$ would be the supremum over all k and θ of the derivative of $p_\theta(x \mid k)$ with respect to k.

The question now is: Can Assumptions 6.5 be relaxed? They probably can, provided that we can prove (as in Lemma 6.14) the a.s. convergence to zero of $\rho(t, \theta)$ and $\eta(t, \theta)$; this result is *the* main thing to prove if we are to follow the NVI approach of Sections 2.4 and 2.5, combined with the estimation of θ using the empirical distribution. A special—but a lot simpler—type of result is discussed in the next section.

2.7 Comments and References

We have presented in this chapter a unified approach to some recent results on adaptive Markov control processes with discounted reward criterion, the unifying theme being the NVI schemes in Section 2.4. We will now make some comments on the related literature and on some possible extensions.

In Sections 2.1 to 2.4, we considered the non-adaptive case. The material in Sections 2.1 and 2.2 (except for Definition 1.3(b) on asymptotic discount optimality) is quite standard: Bertsekas and Shreve (1978), Dynkin and Yushkevich (1979), Hinderer (1970), etc. Many of the basic ideas go back (which doesn't?) to Bellman (1957; 1961), but the DP Theorem 2.2 in its present generality is due to more recent authors, such as Blackwell (1965), Strauch (1966), and Himmelberg et al. (1976). In the proof of Lemma 2.6(b) we followed a recent work by Wakuta (1987). In Section 2.3 we followed Schäl (1981) and Hernández-Lerma (1985a). The most restrictive parts of Assumptions 2.1 on the MCM we used are the boundedness of the one-stage expected reward function $r(x, a)$ and its continuity in $a \in A(x)$. Both conditions can be relaxed but at the expense of complicating the exposition; see, e.g., Bensoussan (1982), Cavazos-Cadena (1986, 1987), Schäl (1981). Another key fact is the contraction property of the DP operator T, which results from the discount factor β being less than 1. If we let β be ≥ 1, we can still get a contraction operator on the space $B(X)$, but with respect to the *span* semi-norm

$$\mathrm{sp}(v) := \sup_x v(x) - \inf_x v(x),$$

and provided that a suitable ergodicity assumption is imposed on the transition law q; see, e.g., Lemma 3.5 in Section 3.3.

The NVI approach in Section 2.4 was introduced by Federgruen and Schweitzer (1981) for Markov decision processes with finite state and control spaces, and first used in adaptive control problems by Hernández-Lerma and Marcus (1985). The finite-state approximations at the end of Section 2.4 were introduced to illustrate another possible use of the NVI schemes. Mandl and Hübner (1985) give another interpretation of the approximating MCM's (X, A, q_t, r_t), where $t = 0, 1, \ldots$. They view these as a nonstationary MCM (Section 1.3) and study the asymptotic normality of the rewards.

Sections 2.5 and 2.6 consider "parametric" and "nonparametric" adaptive MCM's, respectively. The presentation is based (mainly) in Hernández-Lerma (1985a, 1987a) and Hernández-Lerma and Marcus (1987a). The PEC policy was introduced independently by Kurano (1972) and Mandl (1974) for controlled Markov chains with finite state space and *average* reward criterion; Riordon (1969) used similar ideas. For *discounted* reward problems the PEC policy was introduced by Schäl (1981). The modified-PEC policy [Definition 5.15(c)] is a "discounted" version of the policy introduced by Gordienko (1985) for *average* reward problems with unknown disturbance distribution.

Some of the results in this chapter will be extended to average reward problems and to partially observable MCM's in Chapters 3 and 4, respectively, and in Chapter 6 we consider discretization procedures.

There are, of course, other ways to study adaptive control problems. For instance, one can consider the unknown parameter θ as being another state variable and the original problem is then transformed into one with partial state information. This is the typical Bayesian approach; see, e.g., Rieder (1975), Van Hee (1978), Schäl (1979). On the other hand, a particular case that allows many different possibilities is when the state space X is *finite*, for then one can use special techniques, such as learning algorithms (Kurano, 1983; 1987) or numerical approximations (White and Eldeib, 1987), that do not seem to be easily extendable to more general situations. A review of several approaches—mainly for *average* reward problems—is given by Kumar (1985).

Further Remarks on Nonparametric Problems. In the nonparametric adaptive policies in Section 2.6 we used the empirical distribution (cf. 6.9) to estimate the unknown disturbance distribution θ. This approach has advantages and disadvantages. An obvious advantage is that it is very general, in that the disturbance set S and the distribution θ can be "arbitrary", but a disadvantage is that it requires a very restrictive set of assumptions (cf. 6.5) on the control model. However, in the special situation in which S is $\mathbf{R}^d$ and the disturbance distribution θ is absolutely continuous with respect to Lebesgue measure, there are many other ways to estimate θ, some of which are better for the adaptive control problem, in that one can avoid the restrictions imposed by Assumptions 6.5. We will now briefly describe one such situation.

Let us consider again the discrete-time system 6.1, but we suppose now that $S = \mathbf{R}^d$, and that θ has a density $\gamma(y)$. Thus for any Borel set $B \in \mathcal{B}(\mathbf{R}^d)$,

$$\theta(B) = \int_B \gamma(y)dy,$$

and the problem of estimating the disturbance distribution θ becomes a "density estimation" problem, which can be approached in a number of ways [Devroye and Györfi (1985), Hernández-Lerma et al. (1988), Prakasa Rao (1983), Yakowitz (1985), etc.]. We can use, e.g., "kernel" density estimates $\gamma_t(y)$ to estimate $\gamma(y)$, and then the random probability measures defined by

$$\theta_t(B) := \int_B \gamma_t(y)dy, \quad B \in \mathcal{B}(\mathbf{R}^d),$$

are estimates of $\theta(B)$, and, moreover, by Scheffé's Theorem B.3 (Appendix B),

$$\textbf{7.1} \qquad \|\theta_t - \theta\| = \int |\gamma_t(y) - \gamma(t)|dy.$$

We can now obtain 6.4 using a result by Glick (1974) [cf. Devroye and Györfi (1985), p. 10], according to which, as $t \to \infty$, the random variables in 7.1 converge to zero a.s., if

$$\textbf{7.2} \qquad \gamma_t(y) \to \gamma(y) \quad \text{a.s.}$$

for almost all y with respect to Lebesgue measure. Therefore if 7.2 holds, we obtain 6.4, which in turn (as noted in Section 2.6) implies Assumption 5.5, so that the "parametric" results in Section 2.5 can be applied *directly* to the nonparametric problem. This is, roughly, the approach followed by Hernández-Lerma and Duran (1988), who also consider average reward problems, in addition to the discounted case. Alternatively, one can also obtain the same conclusion—namely, the asymptotic discount optimality of the adaptive control policies—combining 7.1 and 7.2 with the Lipschitz property of the function $\theta \to v^*(x,\theta)$ in Remark 5.13, which in the nonparametric case becomes

$$\|v^*(\cdot,\theta) - v^*(\cdot,\theta')\| \leq L_3\|\theta - \theta'\|,$$

where θ and θ' denote disturbance distributions.

A different type of nonparametric adaptive control problem concerns the joint estimation of the disturbance distribution and the regression-like function $F(x,a)$ in systems of the form

7.3 $$x_{t+1} = F(x_t, a_t) + \xi_t.$$

Hernández-Lerma and Doukhan (1988) have studied the special case in which F is of the form $F(x,a) = g(x) + G(x,a)$, where $G(x,a)$ is a given function and $g(x)$ is unknown. Since many applied control models are of the "additive-noise" form 7.3, it would be important to investigate methods to estimate the general, *control-dependent* function $F(x,a)$ in 7.3. In the non-controlled case, i.e., for autoregressive models $x_{t+1} = F(x_t) + \xi_t$, there are well-known techniques to estimate $F(x)$, see, e.g., Doukhan and Ghindès (1983), Schuster and Yakowitz (1979), Yakowitz (1985), Prakasa Rao (1983), or the extensive bibliographical review by Collomb (1981).

3

Average Reward Criterion

3.1 Introduction

Let (X, A, q, r) be a Markov control model (MCM). Thus (by Definition 2.1 in Chapter 1) the state space X and the control set A are Borel spaces; $q(\cdot \mid k)$ is the transition law, a stochastic kernel on X given $\mathbf{K}$, where

$$\mathbf{K} := \{(x, a) \mid x \in X \text{ and } a \in A(x)\},$$

and $r(k)$ is the one-step expected reward function, a real-valued measurable function on $\mathbf{K}$.

We consider in this chapter the problem of maximizing the long-run average expected reward per unit time, or simply, the ***average reward*** defined by

$$\textbf{1.1} \qquad J(\delta, x) := \liminf_{n\to\infty} n^{-1} E_x^\delta \sum_{t=0}^{n-1} r(x_t, a_t)$$

when the initial state is $x_0 = x$ and the policy $\delta \in \Delta$ is used. A policy δ^* is said to be (average-reward) ***optimal*** if it satisfies $J(\delta^*, x) = J^*(x)$ for all $x \in X$, where

$$J^*(x) := \sup_\delta J(\delta, x) \text{ for } x \in X.$$

Actually, under the assumptions imposed below (Sections 3.2 and 3.3) it will follow that J^* is identically constant, say, $J^*(x) = j^*$ for all $x \in X$.

The average reward criterion is ***underselective,*** since it depends "only on the tail of the returns and not on the returns during the first millenium" [Flynn (1976)]. For example, under this criterion, a policy that produces rewards $-R$, say, in periods 0 through n, and R in every period beyond n, for any finite n, is equivalent to (i.e., yields the same average reward as) a policy which has a reward of R in every period. Thus the average reward criterion does not distinguish between these two policies and, therefore, it is said to be underselective. [Several underselective and overselective optimality criteria are discussed, e.g., by Flynn (1976) and Hopp et al. (1986).]

The "disadvantage" of being underselective, however, turns out to be a "good" feature from the point of view of ***adaptive*** control. Indeed, since early performance is not crucial, the "errors" made by the controller during the early periods—when he/she is "learning", through estimates of the true parameter values—are cancelled out in the limit, and therefore, if the controller uses a "consistent" parameter estimation procedure, one would expect the performance to be average-reward optimal. We will show below that this is indeed the case, under appropriate conditions. (As already noted at the beginning of Section 2.3, the situation is different for *discounted* reward problems, in which the best we can hope for, in general, is to obtain *asymptotically* optimal adaptive policies.)

The objective of this chapter is to give a unified presentation to several (some of them, quite recent) results on the approximation and adaptive control of average-reward controlled Markov processes. What follows is an outline of the contents of this chapter.

Summary

Sections 3.2 and 3.3 cover the required background material. In Section 3.2 we obtain some optimality conditions assuming the existence of a bounded solution to the optimality equation (OE) in Theorem 2.2, whereas in Section 3.3 we give several sufficient (ergodicity) conditions for the existence of one such solution.

In Section 3.4 we obtain several uniform approximations to the optimal value function. This section summarizes (and extends to MCM's with Borel state and control spaces) well-known results by several authors on the successive approximations or value-iteration (VI) method.

In Sections 3.5 and 3.6 we consider a sequence of MCM's and give conditions for it to converge to a limit MCM. In both sections, uniform approximations to the optimal value function of the limit MCM, as well as optimal policies, are provided. The main difference to notice between the approximations in Section 3.5 and those in Section 3.6 is that the latter are *recursive,* whereas the former are not.

Section 3.7 deals with adaptive MCM's: the results in Sections 3.4, 3.5 and 3.6 are used to obtain approximations and adaptive policies for MCM's with unknown parameters, provided that a consistent parameter-estimation scheme is given.

We close the chapter in Section 3.8 with general comments on average reward problems and on the related literature.

3.2 The Optimality Equation

The main objective of this section is to prove Theorem 2.2 below, which gives an optimality criterion in terms of the *optimality equation* (OE), provided a solution $\{j^*, v^*(\cdot)\}$ to the (OE) exists. In Section 3.3 we will introduce several conditions insuring the existence of such a solution.

Throughout this chapter, the MCM (X, A, q, r) is supposed to satisfy Assumptions 2.1 below. (These are the same as Assumptions 2.1 in Chapter 2, but they are repeated here for ease of reference.)

2.1 Assumptions.

(a) For each state $x \in X$, $A(x)$ is a (non-empty) compact subset of A.

(b) For some constant R, $|r(k)| \leq R$ for all $k \in \mathbf{K}$, and moreover, for each $x \in X$, $r(x,a)$ is a continuous function of $a \in A(x)$.

(c) $\int v(y)\, q(dy \mid x,a)$ is a continuous function of $a \in A(x)$ for each $x \in X$ and each function $v \in B(X)$.

2.2 Theorem. *Suppose that Assumptions 2.1 hold, and suppose there exists a constant j^* and a function v^* in $B(X)$ such that*

$$j^* + v^*(x) = \max_{a \in A(x)} \left\{ r(x,a) + \int_X v^*(y)\, q(dy \mid x,a) \right\} \quad \text{for all } x \in X. \quad \text{(OE)}$$

Then:

(a) $\sup_\delta J(\delta, x) \leq j^*$ *for all* $x \in X$.

(b) *If $f^* \in \mathbf{F}$ is a stationary policy such that $f^*(x) \in A(x)$ maximizes the right-hand side* (r.h.s.) *of the* (OE), *i.e.*,

$$j^* + v^*(x) = r(x, f^*(x)) + \int v^*(y)\, q(dy \mid x, f^*(x)) \quad \text{for all } x \in X, \quad (1)$$

then f^* is optimal and $J(f^*, x) = j^*$ for all $x \in X$.

(c) *For any policy δ and any $x \in X$,*

$$\lim_{n \to \infty} n^{-1} \sum_{t=0}^{n-1} r(x_t, a_t) = j^* \quad P_x^\delta\text{-a.s.}$$

if and only if

$$\lim_{n \to \infty} n^{-1} \sum_{t=0}^{n-1} \phi(x_t, a_t) = 0 \quad P_x^\delta\text{-a.s.}, \quad (2)$$

where $\phi(x,a)$ is the function defined on $\mathbf{K}$ by

$$\phi(x,a) := r(x,a) + \int v^*(y)\, q(dy \mid x,a) - j^* - v^*(x).$$

(d) *If δ satisfies* (2) *for every $x \in X$, then δ is optimal.*

2.3 Remarks. (a) If j^* and $v^* \in B(X)$ are as in Theorem 2.2, it is then said that $\{j^*, v^*(\cdot)\}$ is a *solution* to the (OE). The (OE) is sometimes called the average-reward Dynamic Programming Equation.

(b) Let T be the operator on $B(X)$ defined, for $v \in B(X)$ and $x \in X$, by

$$Tv(x) := \max_{a \in A(x)} \left\{ r(x,a) + \int v(y)\, q(dy \mid x,a) \right\}.$$

Note that, by Proposition D.3 in Appendix D, $Tv \in B(X)$ whenever $v \in B(X)$, and, on the other hand, we can write the (OE) in terms of T as

$$j^* + v^*(x) = Tv^*(x) \quad \text{for all} \quad x \in X.$$

Note also that, under the assumptions of Theorem 2.2, there exists a stationary policy $f^* \in \mathbf{F}$ that satisfies (1); to see this, use Proposition D.3 again. We will call T the ***dynamic programming*** (DP) operator.

(c) We can also write the (OE) in terms of ϕ:

$$\max_{a \in A(x)} \phi(x,a) = 0 \quad \text{for all} \quad x \in X.$$

Similarly, equation (1) can be written as

$$\phi(x, f^*(x)) = 0 \quad \text{for all} \quad x \in X, \tag{3}$$

and part (b) in Theorem 2.2 can be re-stated as follows: If $f^* \in \mathbf{F}$ is a stationary policy that satisfies (3), then f^* is optimal. Due to these facts and to Theorem 2.2(d), the function ϕ will play in this chapter a similar role to that of the function (also denoted by) ϕ introduced in Section 2.3 to study asymptotic discount optimality.

Proof of Theorem 2.2. To simplify the notation, in this proof we fix an arbitrary policy $\delta \in \Delta$ and an arbitrary initial state $x \in X$, and write E_x^δ and P_x^δ as E and P, respectively.

(a) For any history $h_t = (x_0, a_0, \ldots, x_{t-1}, a_{t-1}, x_t) \in H_t$, it follows from the Markov property 2.5(d) in Chapter 1 that

$$\begin{aligned} E[v^*(x_{t+1}) \mid h_t, a_t] &= \int v^*(y)\, q(dy \mid x_t, a_t) \\ &= r(x_t, a_t) + \int v^*(y)\, q(dy \mid x_t, a_t) - r(x_t, a_t) \\ &\leq j^* + v^*(x_t) - r(x_t, a_t), \quad \text{[by the (OE)]} \end{aligned}$$

and therefore,

$$\sum_{t=0}^{n-1} \{v^*(x_{t+1}) - E[v^*(x_{t+1}) \mid h_t, a_t]\} \geq \sum_{t=0}^{n-1} r(x_t, a_t) + v^*(x_n) - v^*(x_0) - nj^*.$$

Now taking the expectation $E = E_x^\delta$, the expected value of the left-hand side (l.h.s.) is zero, and we obtain

$$nj^* \geq E \sum_{t=0}^{n-1} r(x_t, a_t) + E[v^*(x_n) - v^*(x_0)].$$

Finally, divide by n and the let $n \to \infty$ to obtain $j^* \geq J(\delta, x)$, since v^* is bounded. This proves part (a).

(b) To prove this, it suffices to note that if $f^* \in \mathbf{F}$ satisfies equation (1), then equality holds throughout (with $\delta = f^*$) in the proof of part (a). [We can also conclude (b) from equation (3) and part (d) in the theorem.]

(c) First note that ϕ is bounded. Now, using the Markov property 2.5(d) in Chapter 1 again,

$$\phi(x_t, a_t) = E[r(x_t, a_t) + v^*(x_{t+1}) - v^*(x_t) - j^* \,|\, h_t, a_t],$$

and therefore, substracting $\phi(x_t, a_t)$ on both sides, we get $E(u_t \,|\, h_t, a_t) = 0$, where

$$u_t := r(x_t, a_t) + v^*(x_{t+1}) - v^*(x_t) - j^* - \phi(x_t, a_t).$$

Let $\sigma(h_n)$ be the sigma-algebra generated by h_n. Then $M_n := \sum_{t=0}^{n-1} u_t$, where $n = 1, 2, \ldots$, is a $\sigma(h_n)$-martingale, since M_n is $\sigma(h_n)$-measurable and

$$E(M_{n+1} - M_n \,|\, h_n) = E(u_n \,|\, h_n) = E\{E(u_n \,|\, h_n, a_n) \,|\, h_n\} = 0.$$

Thus, since the u_t are uniformly bounded,

$$\sum_{t=1}^{\infty} t^{-2} E(u_t^2) < \infty,$$

and therefore, by the Strong Law of Large Numbers for martingales (see Lemma 3.11 in Chapter 5),

$$\lim_{n \to \infty} n^{-1} M_n = 0 \quad P\text{-a.s.}$$

By the boundedness of v^*, the latter is equivalent to

$$\lim_{n \to \infty} \left[n^{-1} \sum_{t=0}^{n-1} r(x_t, a_t) - j^* - n^{-1} \sum_{t=0}^{n-1} \phi(x_t, a_t) \right] = 0 \quad P\text{-a.s.},$$

and this in turn implies (c).

(d) This part follows from (c) and the Dominated Convergence Theorem. □

In the proof of Theorem 2.2 notice that the requirement that v^* is bounded can be replaced by the condition that

$$n^{-1} E_x^{\delta} v^*(x_n) \to 0 \quad \text{for every policy } \delta \text{ and every } x \in X.$$

In later sections, Theorem 2.2(d) will be a basic tool to prove optimality. We now turn to the question: Under what conditions does there exist a solution $\{j^*, v^*(\cdot)\}$ to the (OE)?

3.3 Ergodicity Conditions

In this section we introduce several ergodicity conditions that will be useful in the following sections, and which—together with Assumptions 2.1—imply the existence of a solution to the optimality equation (OE) in Theorem 2.2.

3.1 Ergodicity Conditions.

(1) There exists a state $x^* \in X$ and a positive number α_0 such that

$$q(\{x^*\} \mid k) \geq \alpha_0 \quad \text{for all} \quad k \in \mathbf{K}.$$

(2) There exists a measure μ on X such that

$$\mu(X) > 0 \quad \text{and} \quad q(\cdot \mid k) \geq \mu(\cdot) \quad \text{for all} \quad k \in \mathbf{K}.$$

(3) There exists a measure ν on X such that

$$\nu(X) < 2 \quad \text{and} \quad q(\cdot \mid k) \leq \nu(\cdot) \quad \text{for all} \quad k \in \mathbf{K}.$$

(4) There exists a number $\alpha < 1$ such that

$$\sup_{k,k'} \|q(\cdot \mid k) - q(\cdot \mid k')\| \leq 2\alpha,$$

where the sup is over all k and k' in $\mathbf{K}$, and $\| \;\|$ denotes the variation norm for signed measures (Appendix B).

(5) For any stationary policy $f \in \mathbf{F}$ there exists a probability measure p_f on X such that

$$\|q_f^t(\cdot \mid x) - p_f(\cdot)\| \leq c_t \quad \text{for all} \quad x \in X \quad \text{and} \quad t \geq 0,$$

where the numbers c_t are independent of x and f, and $\sum_t c_t < \infty$. Here $q_f^t(\cdot \mid x)$ denotes the t-step transition probability measure of the Markov (state) process $\{x_t\}$ when the stationary policy f is used, given that the initial state is $x_0 = x$. (See Remarks 3.2 below.)

3.2 Remarks. (a) The t-step transition probability $q_f^t(\cdot \mid x) = q^t(\cdot \mid x, f(x))$ in 3.1(5) is given recursively by

$$q_f^t(B \mid x) = \int_X q_f^{t-1}(B \mid y)\, q_f(dy \mid x) \quad \text{for all } B \in \mathcal{B}(X) \text{ and } t \geq 1,$$

where $q_f^0(\cdot \mid x) := p_x(\cdot)$ is the probability measure concentrated at the point $x \in X$, i.e.,

$$p_x(B) := 1_B(x) \quad \text{for all} \quad B \in \mathcal{B}(X).$$

Recall that 1_B denotes the indicator function of set B, and note, on the other hand, that

$$q_f^1(\cdot \mid x) = q_f(\cdot \mid x) = q(\cdot \mid x, f(x)).$$

(b) In 3.1(5), it is easily verified that, for any stationary policy $f \in \mathbf{F}$, p_f is the unique invariant probability measure of the state process $\{x_t\}$, i.e., the unique probability measure satisfying that

$$p_f(B) = \int_X q_f(B \mid x) p_f(dx) \quad \text{for all} \quad B \in \mathcal{B}(X).$$

(c) In the proof of Lemma 3.3 below we will use the following result by Ueno (1957)—also contained in Georgin (1978a): Let $q(\cdot \mid x)$ be the transition probability of a given Markov chain with values in a Borel space X, and let $q^t(\cdot \mid x)$ denote the t-step transition probabilities. Then for every x and y in X,

$$\|q^t(\cdot \mid x) - q^t(\cdot \mid y)\| \leq 2^{-t+1} \sup_{x,y} \|q(\cdot \mid x) - q(\cdot \mid y)\|^t.$$

The number

$$\alpha(q) := 1 - \frac{1}{2} \sup_{x,y} \|q(\cdot \mid x) - q(\cdot \mid y)\|$$

is called the *ergodicity coefficient* [see, e.g., Iosifescu (1972); Ueno (1957)].

3.3 Lemma.

(a) *The following implications hold for the ergodicity conditions* 3.1:

$$\begin{array}{ccccccc} (1) & \rightarrow & (2) & \rightarrow & (4) & \leftarrow & (3) \\ & & & & \downarrow & & \\ & & & & (5) & & \end{array}$$

(b) *For any stationary policy $f \in \mathbf{F}$, any of the ergodicity conditions* 3.1(1) *to* 3.1(5) *implies:*

i. $\lim_{t\to\infty} \int h(y) q_f^t(dy|x) = \int h(y) p_f(dy)$ *for every $h \in B(X)$ uniformly in $x \in X$;*

ii. *the average reward $J(f, x)$ is a constant $j(f)$, that is,*

$$J(f, x) = \int r(y, f(y))\, p_f(dy) =: j(f) \quad \textit{for all} \quad x \in X.$$

Proof. (a) 3.1(1) *implies* 3.1(2). Assume that 3.1(1) holds, and define μ as the measure on X concentrated at x^* with mass α_0, i.e.,

$$\mu(B) := \alpha_0 1_B(x^*) \quad \text{for} \quad B \in B(X).$$

3.1(2) *implies* 3.1(4). Let k and k' be two arbitrary state-action pairs in $\mathbf{K}$ and define the signed measure on X

$$\lambda(\cdot) := q(\cdot \mid k) - q(\cdot \mid k').$$

Then, by the Jordan–Hahn Decomposition Theorem (Appendix B), there exist disjoint measurable sets X^+ and X^- whose union is X and such that

$$\begin{aligned}
\|\lambda\| &= q(X^+ \mid k) - q(X^+ \mid k') - q(X^- \mid k) + q(X^- \mid k') \\
&\leq 1 - \mu(X^+) - \mu(X^-) + 1 \\
&= 2 - \mu(X); \quad \text{take } \alpha = (2 - \mu(X))/2.
\end{aligned}$$

3.1(3) *implies* 3.1(4). With the same notation of the previous proof,

$$\|\lambda\| \leq \nu(X^+) + \nu(X^-) = \nu(X); \quad \text{take } \alpha = \nu(X)/2.$$

3.1(4) *implies* 3.1(5) [Ueno (1957); Georgin (1978a)]. First we will prove that $q_f^t(\cdot \mid x)$ is a Cauchy sequence in variation norm; here we will use results B.1 and B.2 in Appendix B. Since

$$q_f^{t+s}(\cdot \mid x) = \int q_f^t(\cdot \mid y)\, q_f^s(dy \mid x),$$

we obtain

$$\begin{aligned}
\|q_f^t(\cdot \mid x) - q_f^{t+s}(\cdot \mid x)\| &= 2 \sup_B |q_f^t(B \mid x) - q_f^{t+s}(B \mid x)| \quad \text{(by B.2)} \\
&= 2 \sup_B \left| \int \{q_f^t(B \mid x) - q_f^t(B \mid y)\} q_f^s(dy \mid x) \right| \\
&\leq 2 \sup_B \sup_y |q_f^t(B \mid x) - q_f^t(B \mid y)| \quad \text{(by B.1)} \\
&= \sup_y \|q_f^t(\cdot \mid x) - q_f^t(\cdot \mid y)\| \quad \text{(by B.2)} \\
&\leq 2^{-t+1} \sup_{x,y} \|q_f(\cdot \mid x) - q_f(\cdot \mid y)\|^t \text{ (by Remark 3.2(c))} \\
&\leq c_t, \text{ with } c_t := 2\alpha^t. \quad \text{(by 3.1(4))}
\end{aligned}$$

Thus $q_f^t(\cdot \mid x)$ is a Cauchy sequence, and therefore, since the space of finite signed measures endowed with the variation norm is a Banach space, the sequence $q_f^t(\cdot \mid x)$ converges to a (probability) measure p_f on X, which clearly is independent of x. Moreover,

$$\begin{aligned}
\|q_f^t(\cdot \mid x) - p_f(\cdot)\| &\leq \|q_f^t(\cdot \mid x) - q_f^{t+s}(\cdot \mid x)\| + \|q_f^{t+s}(\cdot \mid x) - p_f(\cdot)\| \\
&\leq c_t + \|q_f^{t+s}(\cdot \mid x) - p_f(\cdot)\|,
\end{aligned}$$

and letting s tend to infinity, we conclude 3.1(5). This completes the proof of part (a).

(b) By the implications in part (a), it suffices to—and we will—assume that 3.1(5) holds. Then, for any $f \in \mathbf{F}$, $h \in B(X)$ and $x \in X$,

$$\begin{aligned} \left| \int h(y)\, q_f^t(dy \mid x) \right. &- \left. \int h(y)\, p_f(dy) \right| \\ &\leq \|h\|\, \|q_f^t(\cdot \mid x) - p_f(\cdot)\| \qquad \text{(by B.1)} \\ &\leq \|h\|\, c_t \to 0, \end{aligned}$$

which implies i. To prove ii, note that, as $t \to \infty$,

$$E_x^f r(x_t, a_t) = \int r(y, f(y))\, q_f^t(dy \mid x) \to j(f),$$

so that ii follows from Definition 1.1 of $J(f,x)$. This completes the proof of the lemma. □

To state the following result we introduce some definitions.

3.4 Definitions. The *span* semi-norm of a function $v \in B(X)$ is defined by

$$\operatorname{sp}(v) := \sup_x v(x) - \inf_x v(x).$$

Note that $\operatorname{sp}(v) = 0$ if and only if v is a constant function. Let T be an operator from $B(X)$ into itself. T is said to be a *span-contraction* operator if for some $\alpha \in [0,1)$,

$$\operatorname{sp}(Tu - Tv) \leq \alpha \cdot \operatorname{sp}(u - v) \text{ for every } u \text{ and } v \text{ in } B(X).$$

As in Banach's Fixed Point Theorem (Appendix A), it can be shown that if T is a span-contraction operator on the space $B(X)$, then T has a *span-fixed-point*, i.e., there is a function $v^* \in B(X)$ such that $\operatorname{sp}(Tv^* - v^*) = 0$, or equivalently, $Tv^* - v^*$ is a constant function.

3.5 Lemma. *Suppose that Assumptions* 2.1 *and the ergodicity condition* 3.1(4) *hold. Then the DP operator T in Remark* 2.3(b) *is a span-contraction operator, i.e.,*

$$\operatorname{sp}(Tu_1 - Tu_2) \leq \alpha \cdot \operatorname{sp}(u_1 - u_2) \textit{ for every } u_1 \textit{and } u_2 \textit{ in } B(X),$$

where $\alpha < 1$ is the number in 3.1(4).

Proof. First we shall prove the following: For any two state-action pairs $k = (x,a)$ and $k' = (x', a')$ in $\mathbf{K}$, and any function $v \in B(X)$,

$$\int v(y)\, q(dy \mid k) - \int v(y)\, q(dy \mid k') = \int v(y)\, \lambda(dy) \leq \alpha \cdot \operatorname{sp}(v), \qquad (*)$$

where λ is the signed measure on X defined by

$$\lambda(\cdot) := q(\cdot \mid k) - q(\cdot \mid k').$$

Indeed, by the Jordan–Hahn Decomposition Theorem (Appendix B), there exist disjoint measurable sets X^+ and X^- whose union is X and such that

$$\|\lambda\| = \lambda(X^+) - \lambda(X^-) \leq 2\alpha,$$

where the latter inequality comes from 3.1(4). On the other hand, since

$$\lambda(X) = \lambda(X^+) + \lambda(X^-) = 0,$$

we have $\lambda(X^+) \leq \alpha$. Finally we can write the integral on the l.h.s. of (*) as

$$\begin{aligned}
\int v(y)\,\lambda(dy) &= \int_{X+} v\,d\lambda + \int_{X-} v\,d\lambda \\
&\leq \int_{X+} \left(\sup_y v(y)\right) d\lambda + \int_{X-} \left(\inf_y v(y)\right) d\lambda \\
&\leq \lambda(X^+)\cdot \mathrm{sp}(v) + \left(\inf_y v(y)\right)\cdot \lambda(X) \\
&\leq \alpha \cdot \mathrm{sp}(v),
\end{aligned}$$

which proves (*).

Now let u_1 and u_2 be arbitrary functions in $B(X)$, and (using Proposition D.3 in Appendix D) let $g_1 \in \mathbf{F}$ and $g_2 \in \mathbf{F}$ be such that

$$Tu_i(x) = r(x, g_i(x)) + \int u_i(y)\,q(dy \mid x, g_i(x)) \quad \text{for all } x \in X \text{ and } i = 1, 2.$$

Of course, if $i \neq j$, then

$$Tu_i(x) \geq r(x, g_j(x)) + \int u_i(y)\,q(dy \mid x, g_j(x)),$$

and therefore, for any two states x and x' in X,

$$(Tu_1 - Tu_2)(x) - (Tu_1 - Tu_2)(x') \leq \int [u_1(y) - u_2(y)]\lambda(dy)$$

[where λ is the signed measure in (*), with $k = (x, g_1(x))$ and $k' = (x', g_2(x'))$]

$$\leq \alpha \cdot \mathrm{sp}(u_1 - u_2).$$

Since x and x' are arbitrary, the desired result follows. □

It follows from Lemma 3.5 that T has a span-fixed-point $v^* \in B(X)$; that is, there exists $v^* \in B(X)$ and a constant j^* such that

$$Tv^*(x) - v^*(x) = j^* \quad \text{for all } x \in X,$$

which is the same as the optimality equation (OE) in Remark 2.3(b). Therefore, from Lemmas 3.3(a) and 3.5 we conclude the following.

3.6 Corollary. *Suppose that Assumptions* 2.1 *and any of the ergodicity conditions* 3.1(1) *to* 3.1(4) *hold. Then there exists a solution* $\{j^*, v^*(\cdot)\}$ *to the optimality equation* (*OE*).

3.7 Comments. The idea of using an ergodicity condition to obtain a span-contraction operator has been used by many authors: Tijms (1975), Hübner (1977), Rieder (1979), etc. If instead of using 3.1(4), as in Lemma 3.5, we use directly 3.1(1), or 3.1(2), or 3.1(3), the proofs are simpler. For instance, suppose that 3.1(2) holds and let $q'(\cdot \mid k)$ be the measure on X defined by

$$q'(\cdot \mid k) := q(\cdot \mid k) - \mu(\cdot) \quad \text{for} \quad k \in \mathbf{K}.$$

Then it is easy to see (using either Proposition A.2 or Proposition A.3 in Appendix A) that the operator T' on $B(X)$ defined by

$$\begin{aligned} T'v(x) &:= \max_{a \in A(x)} \left[r(x,a) + \int v(y)\, q'(dy \mid x, a) \right] \\ &= Tv(x) - \int v(y)\, \mu(dy) \end{aligned}$$

is a contraction on $B(X)$; in fact,

$$\|T'u - T'v\| \leq \beta \|u - v\|, \quad \text{where} \quad \beta := 1 - \mu(X) < 1.$$

Thus T' has a unique fixed point $v^* \in B(X)$:

$$v^* = T'v^* = Tv^* - \int v^*(y)\, \mu(dy),$$

which, taking $j^* := \int v^*(y)\, \mu(dy)$, is the same as the (OE) $v^* = Tv^* - j^*$. A similar conclusion is obtained if now 3.1(3) holds and we take on X the measure

$$q''(\cdot \mid k) := \nu(\cdot) - q(\cdot \mid k), \quad \text{where} \quad k \in \mathbf{K}.$$

The latter approach has been used by Gubenko and Statland (1975), Kurano (1985, 1986), Ross (1968), etc. A "minorization condition" such as 3.1(2) is often used in the study of ergodicity of Markov chains [see, e.g., Nummelin and Tuominen (1982)].

Alternatively, to obtain the conclusion of Corollary 3.6 we can use the ergodicity condition 3.1(5) directly and obtain a solution $\{j^*, v^*(\cdot)\}$ to the (OE) as a limit of the form

$$j^* = \lim(1-\beta)v_\beta(z) \quad \text{and} \quad v^*(x) = \lim[v_\beta(x) - v_\beta(z)],$$

where the limit is with respect to an increasing sequence $\beta = \beta_n$ of positive numbers converging to 1; v_β is the optimal reward function of β-discounted MCM's (Section 2.2) and $z \in X$ is a *fixed* arbitrary state. [See, e.g., Georgin (1978a), Gordienko (1985), Ross (1968),... .]

We now turn to the question of how to obtain (uniform) approximations to the optimal average reward j^*.

3.4 Value Iteration

A useful approach to obtain uniform approximations to j^* is to use the method of successive approximations or ***value iteration*** (VI)—which we have already seen for ***discounted*** reward problems in Section 2.2.

Throughout the remainder part of this chapter we suppose that the MCM (X, A, q, r) satisfies the following.

4.1 Assumption. Assumptions 2.1 *and* the ergodicity condition 3.1(4) hold.

Thus, by Lemma 3.5 and Corollary 3.6, the DP operator T is a span-contraction operator and there exists a bounded solution $\{j^*, v^*(\cdot)\}$ to the (OE)

4.2 $$j^* + v^*(x) = Tv^*(x) \quad \text{for all } \; x \in X.$$

Let us now define the ***value iteration*** (VI) functions $v_t \in B(X)$ by

$$v_t := Tv_{t-1} = T^t v_0, \quad \text{for } \; t = 1, 2, \ldots,$$

where $v_0 \in B(X)$ is arbitrary; that is, for $t \geq 1$ and $x \in X$,

4.3 $$v_t(x) := \max_{a \in A(x)} \left\{ r(x, a) + \int_X v_{t-1}(y)\, q(dy \mid x, a) \right\}.$$

$v_t(x)$ can be interpreted as the maximal expected reward for a planning horizon of t epochs when the initial state is $x_0 = x$ and the terminal reward $v_0(y)$ is obtained if the final state is $x_t = y$; that is, for any $t \geq 1$ and $x \in X$,

$$v_t(x) = \sup_{\delta \in \Delta} E_x^\delta \left\{ \sum_{i=0}^{t-1} r(x_i, a_i) + v_0(x_t) \right\}.$$

This is a standard result in (finite-horizon) dynamic programming [Bertsekas (1976; 1987), Kushner (1971), etc.], and it can be proved using the definition of v_t and the Markov properties in Remark 2.5, Chapter 1. Clearly, as $t \to \infty$, v_t might not converge to a function in $B(X)$: take, for instance, $r(x, a)$ identical to a non-zero constant. We shall see, however, that appropriate transformations of v_t do converge.

Uniform Approximations

Let us define a sequence of functions e_t in $B(X)$ by

4.4 $e_t(x) := T^t v_0(x) - T^t v^*(x) = v_t(x) - v^*(x) - tj^*$ for every $t \geq 0$ and $x \in X$,

where the second equality follows from 4.2, since it implies $T^t v^* = v^* + tj^*$ for every $t \geq 0$. We also note that, for every $t \geq 0$ and $x \in X$,

4.5
$$e_{t+1}(x) = \max_{a\in A(x)} \left\{ \phi(x,a) + \int e_t(y)\, q(dy \mid x,a) \right\},$$

where $\phi(x,a)$ is the function on $\mathbf{K}$ defined in Theorem 2.2(c). Other properties of e_t are collected in the following lemma.

4.6 Lemma. *Suppose that Assumption* 4.1 *holds and let* $\alpha < 1$ *be the number in the ergodicity condition* 3.1(4). *Then:*

(a) $\text{sp}(e_t) \leq \alpha \cdot \text{sp}(e_{t-1}) \leq \alpha^t \cdot \text{sp}(e_0)$ *for all* $t \geq 0$.

(b) *The sequence* $e_t^+ := \sup_x e_t(x)$ *is nonincreasing, whereas* $e_t^- := \inf_x e_t(x)$ *is nondecreasing*

(c) $\sup_x |e_t(x) - c| \leq \alpha^t \cdot \text{sp}(e_0)$ *for all* t, *where* $c = \lim e_t^+ = \lim e_t^-$.

(d) $\|e_t\| \leq \|e_{t-1}\| \leq \|e_0\|$ *for all* t.

Proof. (a) This part follows from the span-contraction property of T (Lemma 3.5):

$$\text{sp}(e_t) = \text{sp}(T^t v_0 - T^t v^*) \leq \alpha \cdot \text{sp}(T^{t-1} v_0 - T^{t-1} v^*) = \alpha \cdot \text{sp}(e_{t-1}).$$

(b) Since $\phi(x,a) \leq 0$, 4.5 implies

$$e_{t+1}(x) \leq \sup_y e_t(y) = e_t^+,$$

so that $e_{t+1}^+ \leq e_t^+$. To prove the second part, let $f \in \mathbf{F}$ be a stationary policy such that $\phi(x, f(x)) = 0$ for all $x \in X$ [see Remarks 2.3(b) and (c)]. Then, from 4.5,

$$e_{t+1}(x) \geq \int e_t(y)\, q_f(dy \mid x) \geq e_t^-.$$

(c) This part follows from (a) and (b), since $\text{sp}(e_t) = e_t^+ - e_t^-$.

(d) Using again 4.5 and $\phi(x,a) \leq 0$, we obtain $|e_{t+1}(x)| \leq \|e_t\|$ for all $t \geq 0$ and $x \in X$, which implies the result. □

By Lemma 4.6, the sequence $\{e_t\}$ has nice properties: it is uniformly bounded, it decreases in both the sup norm and the span semi-norm, and it converges exponentially fast to a constant. We will use these properties in Theorem 4.8 below, but first let us introduce a definition.

4.7 Definition. Let $\delta = \{f_t\}$ be a Markov policy such that $f_t(x) \in A(x)$ maximizes the r.h.s. of 4.3 for each $t \geq 1$ and $x \in X$, that is,

$$v_t(x) = r(x, f_t(x)) + \int v_{t-1}(y)\, q(dy \mid x, f_t(x)); \tag{1}$$

we take $f_0 \in \mathbf{F}$ arbitrary. We call δ a *value-iteration* (VI) policy.

4.8 Theorem. *Suppose that Assumption 4.1 holds and let $\alpha < 1$ be as 3.1(4). Let V_t^+ and V_t^- be the sequences defined by*

$$V_t^+ := \sup_x w_t(x) \quad \text{and} \quad V_t^- := \inf_x w_t(x),$$

where $w_t(x) := v_t(x) - v_{t-1}(x)$ for $t \geq 1$ and $x \in X$. Then:

(a) *The sequence V_t^+ is nonincreasing, V_t^- is nondecreasing, and both sequences converge exponentially fast to j^*; namely, for all $t \geq 1$,*

$$-\alpha^{t-1}\text{sp}(e_0) \leq V_t^- - j^* \leq V_t^+ - j^* \leq \alpha^{t-1}\text{sp}(e_0).$$

(b) $V_t^- \leq J(f_t, x) \leq j^* \leq V_t^+$ *for all $t \geq 1$ and $x \in X$, so that, from part* (a),

$$\sup_x |J(f_t, x) - j^*| \leq \alpha^{t-1} \cdot \text{sp}(e_0) \quad \text{for all } t \geq 1,$$

where $\delta = \{f_t\}$ is the VI policy in Definition 4.7.

(c) $\sup_x |J(f_t, x) - j^*| \leq \sup_x |w_t(x) - j^*| \leq \alpha^{t-1} \cdot \text{sp}(e_0)$ *for all $t \geq 1$.*

(d) *For every fixed $z \in X$,* $\sup_x |[v_t(x) - v_t(z)] - [v^*(x) - v^*(z)]| \leq 2\alpha^t \cdot \text{sp}(e_0)$ *for all $t \geq 0$.*

(e) $\sup_x |\phi(x, f_t(x))| \leq 2\alpha^{t-1} \cdot \text{sp}(e_0) \to 0$ *as $t \to \infty$, and therefore, by Theorem* 2.2(d), *the VI policy $\delta = \{f_t\}$, i.e., the policy that uses the control $a_t := f_t(x_t)$ at time t, is optimal.*

This theorem provides several uniform approximations to the optimal average reward j^*. Note also that, by parts (b), (c) and (e), the stationary policy $f_t \in \mathbf{F}$ may be regarded as "approximately" optimal for the infinite horizon problem when t is sufficiently large. The "relative value functions" $v_t(x) - v_t(z)$ in part (d) were introduced by D.J. White (1963).

Proof. (a) By Definition 4.7 of $f_t \in \mathbf{F}$,

$$v_t(x) = r(x, f_t(x)) + \int v_{t-1}(y)\, q(dy \mid x, f_t(x)),$$

whereas

$$v_{t-1}(x) \geq r(x, f_t(x)) + \int v_{t-2}(y)\, q(dy \mid x, f_t(x)).$$

Therefore,

$$w_t(x) \leq \int w_{t-1}(y)\, q(dy \mid x, f_t(x)) \leq V_{t-1}^+,$$

which implies $V_t^+ \leq V_{t-1}^+$. A similar argument yields $V_t^- \geq V_{t-1}^-$. This proves the first part of (a).

To prove the second part, note that, by the definition 4.4 of e_t,

$$\begin{aligned} w_t(x) &= v_t(x) - v_{t-1}(x) \\ &= e_t(x) - e_{t-1}(x) + j^*, \end{aligned}$$

so that

$$\begin{aligned} V_t^+ &= \sup_x \{e_t(x) - e_{t-1}(x)\} + j^* \\ &\leq e_t^+ - e_{t-1}^- + j^* \\ &\leq \mathrm{sp}(e_{t-1}) + j^* \qquad \text{[by Lemma 4.6(b)]} \\ &\leq \alpha^{t-1} \cdot \mathrm{sp}(e_0) + j^*, \qquad \text{[by Lemma 4.6(a)]} \end{aligned}$$

and similarly,

$$V_t^- = \inf_x \{e_t(x) - e_{t-1}(x)\} + j^* \geq -\alpha^{t-1} \cdot \mathrm{sp}(e_0) + j^*.$$

(b) It follows from part (a) that $V_t^- \leq j^* \leq V_t^+$ for all t. On the other hand, $J(f_t, x) \leq j^*$ for all $t \geq 0$ and $x \in X$. Thus it only remains to prove the first inequality in (b).

To prove this, let us simplify the notation as follows: we fix any integer $t \geq 1$ and write f_t as f, so that equation (1) in Definition 4.7 becomes

$$v_t(x) = r(x, f(x)) + \int v_{t-1}(y)\, q_f(dy \mid x).$$

Now integrating both sides with respect to the invariant probability measure p_f, it follows from Remark 3.2(b) and Lemma 3.3(b.ii) that

$$\begin{aligned} \int v_t(x)\, p_f(dx) &= \int r(x, f(x))\, p_f(dx) + \int\int v_{t-1}(y)\, q_f(dy \mid x)\, p_f(dx) \\ &= j(f) + \int v_{t-1}(y)\, p_f(dy). \end{aligned}$$

Therefore,

$$j(f) = \int w_t(y)\, p_f(dy), \quad \text{with} \quad f = f_t, \tag{2}$$

which implies $V_t^- \leq j(f_t) \leq V_t^+$ for all $t \geq 1$, where $j(f_t) = J(f_t, x)$ for all x.

(c) The first inequality follows from equation (2), and the second follows from part (a), since $V_t^- \leq w_t(x) \leq V_t^+$ for all $t \geq 1$ and $x \in X$.

(d) This follows from Definition 4.4 of e_t and Lemma 4.6(c).

(e) From 4.5 and equation (1) in Definition 4.7,

$$\phi(x, f_t(x)) = e_t(x) - \int e_{t-1}(y)\, q(dy \mid x, f_t(x)),$$

so that, by Lemma 4.6(c),

$$\begin{aligned}\sup_x |\phi(x, f_t(x))| &\leq \sup_x |e_t(x) - c| + \sup_x |e_{t-1}(x) - c| \\ &\leq 2\alpha^{t-1} \cdot \mathrm{sp}(e_0).\end{aligned}$$

This completes the proof of Theorem 4.8. □

Successive Averagings

As a direct consequence of the previous results we will now extend to MCM's with Borel state and control spaces a theorem by Baranov (1982) on successive averagings for MCM's with finitely many states and controls.

Let v_t be the VI functions in 4.3 and define $u_t := t^{-1}v_t$ for all $t \geq 1$, and $u_0 := 0$. Using 4.3 we can write the u_t iteratively:

4.9 $$u_t = Q_t u_{t-1} \quad \text{for all } t \geq 1,$$

where Q_t is the operator on $B(X)$ given, for every $v \in B(X)$ and $x \in X$, by

$$Q_t v(x) := \max_{a \in A(x)} \left\{ t^{-1} r(x, a) + (t-1)t^{-1} \int v(y)\, q(dy \mid x, a) \right\}.$$

For each $t \geq 1$, Q_t is a contraction operator with modulus $(t-1)/t$, and therefore, there exists a unique function u_t^* in $B(X)$ such that

4.10 $$u_t^* = Q_t u_t^*.$$

From Lemma 4.6 and Theorem 4.8, we then obtain the following.

4.11 Corollary. *Suppose that the assumptions of Theorem* 4.8 *hold. Then, as* $t \to \infty$, *each of the following converges to zero:*

(a) $\sup_x |u_t(x) - j^*|$,

(b) $\|u_t^* - u_t\|$, *and*

(c) $\sup_x |u_t^*(x) - j^*|$.

Proof. Part (a) follows from Lemma 4.6(c), whereas part (b) follows from:

$$\begin{aligned}\|u_t^* - u_t\| &\leq (t-1)\|u_t - u_{t-1}\| = \|w_t - u_t\| \\ &\leq \sup_x |w_t(x) - j^*| + \sup_x |u_t(x) - j^*|,\end{aligned}$$

where the w_t are the functions defined in Theorem 4.8.

Finally, part (c) follows from (a) and (b). □

4.12 Remark. Observe that the VI policy $\delta = \{f_t\}$ in Definition 4.7 is such that $f_t(x) \in A(x)$ also maximizes the r.h.s. of the "successive averagings" equation 4.9. Thus, from Theorem 4.8(e), we have obtained by a different approach another conclusion in Baranov's (1982) paper: The policy δ defined via de equations 4.9 is optimal. It can also be proved, using Corollary 4.11(b) and (c), that the Markov policy $\delta' = \{f_t'\}$ such that $f_t'(x)$ maximizes the r.h.s. of 4.10 is optimal.

3.5 Approximating Models

Let (X, A, q_t, r_t), where $t = 0, 1, \ldots$, be a sequence of MCM's. In this section we give conditions under which the average-optimal reward of the t-model converges to the optimal reward of a "limit" MCM (X, A, q, r); in the following section we will use the t-models to study the convergence of a nonstationary version of the value iteration functions in Section 3.4.

In the limiting MCM (X, A, q, r), sometimes we will write the transition law q and the one-step reward r as q_∞ and r_∞, respectively. We will thus write statements valid for all $0 \le t \le \infty$.

Throughout this section we suppose the following.

5.1 Assumptions. For each $0 \le t \le \infty$, the t-MCM (X, A, q_t, r_t) satisfies Assumptions 4.1. Moreover, the sequence $\{r_t\}$ is uniformly bounded and 3.1(4) holds uniformly in t; that is,

(a) $|r_t(k)| \le R < \infty$ for all $k \in \mathbf{K}$ and $0 \le t \le \infty$, and

(b) $\sup_{t,k,k'} \|q_t(\cdot \mid k) - q_t(\cdot \mid k')\| \le 2\alpha$, with $\alpha < 1$, where the sup is over all k and k' in $\mathbf{K}$ and all $0 \le t \le \infty$.

In addition:

(c) The sequences $\rho(t)$ and $\pi(t)$ converge to zero as $t \to \infty$, where

$$\rho(t) := \sup_k |r_t(k) - r(k)| \quad \text{and} \quad \pi(t) := \sup_k \|q_t(\cdot \mid k) - q(\cdot \mid k)\|,$$

and the sup is over all k in $\mathbf{K}$.

Thus, for each t, all the results in Sections 3.2, 3.3 and 3.4 hold. In particular, by Corollary 3.6, for each t, there exists a bounded solution $\{j_t^*, v_t^*(\cdot)\}$ to the optimality equation for the t-MCM:

$$\textbf{5.2} \qquad j_t^* + v_t^*(x) = \max_{a \in A(x)} \left\{ r_t(x,a) + \int v_t^*(y)\, q_t(dy \mid x, a) \right\}$$
$$= T_t v_t^*(x) \quad \text{for all } x \in X \text{ and } 0 \le t \le \infty,$$

where T_t is the operator on $B(X)$ defined, for $v \in B(X)$ and $x \in X$, by

$$\textbf{5.3} \qquad T_t v(x) := \max_{a \in A(x)} \left\{ r_t(x,a) + \int v(y)\, q_t(dy \mid x, a) \right\}.$$

For the limit MCM $(X, A, q, r) = (X, A, q_\infty, r_\infty)$, we write (sometimes) $j^* = j_\infty^*$ and $v^* = v_\infty^*$, so that 5.2 and 5.3 hold for $t = \infty$.

We will also use the following *notation:* for each t, we denote by

$$q_{t,f}^n(\cdot \mid x)$$

the n-step transition probability for the t-model when the stationary policy $f \in \mathbf{F}$ is used, given that the initial state is $x_0 = x$. We also write

$$q^n_{\infty,f}(\cdot \mid x) \quad \text{as} \quad q^n_f(\cdot \mid x);$$

see Remark 3.2(a). Then, we can re-state Lemmas 3.3 and 3.5 in combined form as follows.

5.4 Lemma. *Suppose that Assumptions* 5.1 *hold.*

(a) 5.1(b) *implies that for each* $0 \le t \le \infty$ *and any stationary policy* $f \in \mathbf{F}$, *there exists a probability measure* $p_{t,f}$ *on* X *such that*

$$\|q^n_{t,f}(\cdot \mid x) - p_{t,f}(\cdot)\| \le c_n \quad \textit{for all } x \in X \textit{ and } n \ge 0,$$

where the numbers c_n *are independent of* t, f, *and* x, *and* $\sum c_n < \infty$. *(As in the proof of Lemma* 3.3(a), *we can take* $c_n = 2\alpha^n$.*) We write* $p_{\infty,f}$ *as* p_f.

(b) *Let* $J_t(f, x)$ *be the average reward for the t-MCM when the stationary policy* f *is used and* $x_0 = x$. *Then* $J_t(f, x)$ *is a constant* $j_t(f)$ *in* x:

$$\begin{aligned} J_t(f,x) &:= \liminf_n n^{-1} \sum_{i=0}^{n-1} E^f_x r_t(x_i, a_i) \\ &= \lim_n n^{-1} \sum_{i=0}^{n-1} \int r_t(y, f(y))\, q^i_{t,f}(dy \mid x) \\ &= \lim_n \int r_t(y, f(y))\, q^n_{t,f}(dy \mid x) \qquad (1) \\ &= \int r_t(y, f(y))\, p_{t,f}(dy) \\ &=: j_t(f) \quad \textit{for all } x \in X. \end{aligned}$$

(c) *For each* t, *the operator* T_t *in* 5.3 *is a span-contraction:*

$$\text{sp}(T_t u - T_t v) \le \alpha \cdot \text{sp}(u - v) \quad \textit{for every } u \textit{ and } v \textit{ in } B(X).$$

In addition we have:

(d) *For every* $0 \le t \le \infty$, $f \in \mathbf{F}$ *and* $x \in X$,

$$\|q^n_{t,f}(\cdot \mid x) - q^n_f(\cdot \mid x)\| \le n\pi(t) \quad \textit{for all } n = 0, 1, \ldots.$$

(e) *As* $t \to \infty$, $\|p_{t,f}(\cdot) - p_f(\cdot)\| \to 0$ *uniformly in* $f \in \mathbf{F}$.

(f) *As* $t \to \infty$, $|j_t(f) - j(f)| \to 0$ *uniformly in* f, *where* $j(f) = j_\infty(f)$.

Proof. Parts (a) and (b) come from Lemma 3.3, and (c) comes from Lemma 3.5.

(d) For $n = 0$, part (d) holds trivially, since $q^0_{t,f}(\cdot \mid x)$ is the probability measure concentrated at x for every t, f and x; see Remark 3.2(a). For $n = 1$, the inequality follows from the definition of $\pi(t)$ in 5.1(c). For $n > 1$, the inequality in (d) is easily verified by induction; use Remark 3.2(a).

(e) This part follows from (a) and (d) using the inequality

$$\begin{aligned} \|p_{t,f} - p_f\| &\leq \|q^n_{t,f}(\cdot \mid x) - p_{t,f}(\cdot)\| + \|q^n_f(\cdot \mid x) - p_f(\cdot)\| \\ &\quad + \|q^n_{t,f}(\cdot \mid x) - q^n_f(\cdot \mid x)\| \\ &\leq 2\, c_n + n\pi(t). \end{aligned}$$

(f) From equality (1) in part (b), we obtain

$$\begin{aligned} |j_t(f) - j(f)| &= \left| \int r_t(y, f(y)) p_{t,f}(dy) - \int r(y, f(y)) p_f(dy) \right| \\ &\leq \rho(t) + R\|p_{t,f}(\cdot) - p_f(\cdot)\|, \end{aligned}$$

which implies the result. [Alternatively, (f) follows from (e), using Proposition B.4 in Appendix B.] This completes the proof of the lemma. □

We now introduce a policy that will be shown to be optimal for the limiting MCM, and that will be used in Section 3.7 to define the "principle of estimation and control (PEC)" adaptive policy.

5.5 Definition. For each t, let $f^*_t \in \mathbf{F}$ be such that $f^*_t(x) \in A(x)$ maximizes the r.h.s. of 5.2 for every $x \in X$, and let $\delta^* = \{f^*_t\}$ be the Markov policy for the MCM (X, A, q, r) that takes the action

$$a_t := f^*_t(x_t) \quad \text{at time} \quad t = 0, 1, \ldots .$$

5.6 Remarks. For each t, Theorem 2.2(b) implies that $f^*_t \in \mathbf{F}$ is an optimal stationary policy for the t-model (X, A, q_t, r_t), so that $J_t(f^*_t, x) = j_t(f^*_t) = j^*_t$. Note also that, by the definition of f^*_t,

$$\phi_t(x, f^*_t(x)) = 0 \quad \text{for all} \quad x \in X, \tag{2}$$

where ϕ_t is the function on $\mathbf{K}$ defined by

$$\phi_t(x, a) := r_t(x, a) + \int v^*_t(y)\, q_t(dy \mid x, a) - j^*_t - v^*_t(x);$$

cf. Remark 2.3(c).

We are now ready for the main result in this section.

5.7 Theorem. *Suppose that Assumptions* 5.1 *hold. Then:*

(a) $\lim_{t\to\infty} |j_t^* - j^*| = 0$.

(b) $\mathrm{sp}(v_t^* - v^*) \le b_0 \cdot \max\{\rho(t), \pi(t)\}$, *where* $b_0 := 2(1 + \|v^*\|)/(1-\alpha)$.

(c) *The Markov policy* $\delta^* = \{f_t^*\}$ *in Definition* 5.5 *is average-optimal for the MCM* (X, A, q, r).

Proof. (a) Since

$$j_t^* = \sup_f j_t(f) \quad \text{for all} \quad 0 \le t \le \infty \quad (\text{with } j_\infty^* = j^*),$$

where the sup is over all $f \in \mathbf{F}$, Proposition A.3 in Appendix A yields

$$|j_t^* - j^*| \le \sup_f |j_t(f) - j(f)|.$$

Thus part (a) follows from Lemma 5.4(f).

(b) By the span-contraction property of T_t (Lemma 5.4(c)),

$$\mathrm{sp}(T_t v_t^* - T_t v^*) \le \alpha \cdot \mathrm{sp}(v_t^* - v^*),$$

and, on the other hand, a straightforward calculation yields

$$\begin{aligned} \mathrm{sp}(T_t v^* - T v^*) &\le 2\|T_t v^* - T v^*\| \\ &\le 2[\rho(t) + \|v^*\|\,\pi(t)] \\ &\le 2(1 + \|v^*\|) \cdot \max\{\rho(t), \pi(t)\}. \end{aligned}$$

Now, by 5.2, $v_t^* = T_t v_t^* - j_t^*$ for all $0 \le t \le \infty$, so that

$$\begin{aligned} \mathrm{sp}(v_t^* - v^*) &= \mathrm{sp}(T_t v_t^* - T v^*) \\ &\le \mathrm{sp}(T_t v_t^* - T_t v^*) + \mathrm{sp}(T_t v^* - T v^*) \\ &\le \alpha \cdot \mathrm{sp}(v_t^* - v^*) + 2(1 + \|v^*\|) \cdot \max\{\rho(t), \pi(t)\}, \end{aligned}$$

which is equivalent to (b).

(c) By Theorem 2.2(d), to prove (c) it suffices to show that, as $t \to \infty$,

$$\sup_x |\phi(x, f_t^*(x))| \to 0. \tag{3}$$

Now, by (2) in Remark 5.6, we can write

$$\phi(x, f_t^*(x)) = \phi(x, f_t^*(x)) - \phi_t(x, f_t^*(x))$$

and expanding the r.h.s. using the definitions of ϕ and ϕ_t, a direct calculation shows that

$$|\phi(x, f_t^*(x))| \le \rho(t) + \|v^*\|\,\pi(t) + |j_t^* - j^*| + \mathrm{sp}(v_t^* - v^*),$$

and therefore (3) follows from parts (a) and (b) and Assumption 5.1(c). This completes the proof of the theorem. □

The results in Theorem 5.7 are, of course, very nice, but they have an inconvenience: to use any of parts (a), (b) or (c), first we have to solve the optimality equation 5.2 for each $t = 0, 1 \ldots$. Certainly, it would be nicer, to have a *recursive* approximation scheme. We will do this in the following section.

3.6 Nonstationary Value Iteration

We shall consider again the sequence of MCM's (X, A, q_t, r_t) in Section 3.5, which "converge" to a limit MCM $(X, A, q, r) = (X, A, q_\infty, r_\infty)$. Now, however, we impose assumptions more restrictive than 5.1. These are the following.

6.1 Assumptions. In addition to Assumptions 5.1, we suppose that

$$\overline{\rho} := \sum_{t=0}^{\infty} \rho(t) < \infty \quad \text{and} \quad \overline{\pi} := \sum_{t=0}^{\infty} \pi(t) < \infty,$$

where $\rho(t)$ and $\pi(t)$ are the sequences defined 5.1(c).

The necessity of the new assumptions when doing ***nonstationary value iteration*** (NVI) is discussed by Federgruen and Schweitzer (1981, pp. 232–233), who introduced the NVI scheme for Markov decision processes with *finite* state and control spaces. In this section we will extend the NVI scheme to MCM's with Borel state and control spaces satisfying Assumptions 6.1; we will follow a development somewhat parallel to that in Section 3.4.

Thus, instead of the VI functions v_t in 4.3, we now define recursively the NVI functions $\overline{v}_t$ as follows: Let $\overline{v}_0$ be an arbitrary function in $B(X)$, and for every $t \geq 0$ and $x \in X$, let

$$\textbf{6.2} \qquad \overline{v}_{t+1}(x) := T_t \overline{v}_t(x) = \max_{a \in A(x)} \left\{ r_t(x, a) + \int \overline{v}_t(y)\, q_t(dy \mid x, a) \right\},$$

where the second equality comes from Definition 5.3 of the operator T_t. And also, instead of the functions e_t in 4.4, we now introduce the functions d_t defined below.

6.3 Definition. For each $t \geq 0$ and $x \in X$, let

$$d_t(x) := \overline{v}_t(x) - T^t v^*(x) = \overline{v}_t(x) - tj^* - v^*(x),$$

and

$$c_t(x) := d_t(x) + v^*(x) = \overline{v}_t(x) - tj^*.$$

We write the span seminorm of d_t as $\text{sp}(d_t) = d_t^+ - d_t^-$, where

$$d_t^+ := \sup_x d_t(x) \quad \text{and} \quad d_t^- := \inf_x d_t(x).$$

We will first prove that the functions d_t, and therefore, the c_t, are uniformly bounded in the sup norm.

6.4 Lemma. *There exists a constant M such that* $\text{sp}(d_t) \leq 2\|d_t\| \leq M$ *and* $\|c_t\| \leq M$ *for all* t.

Proof. By Definitions 6.2 and 6.3 and the (OE) $v^* = Tv^* - j^*$ in Theorem 2.2,

$$\begin{aligned} d_{t+1}(x) &= T_t\overline{v}_t(x) - (t+1)j^* - v^*(x) \\ &= \max_{a\in A(x)} \{r_t(x,a) - r(x,a) + \int d_t(y)\, q_t(dy \mid x,a) + \phi(x,a) \\ &\quad + \int v^*(y)[q_t(dy \mid x,a) - q(dy \mid x,a)]\}, \end{aligned}$$

where $\phi(x,a)$ is the function defined in Theorem 2.2(c). Therefore, since $\phi(x,a) \leq 0$, we obtain, for all t,

$$d_{t+1}(x) \leq \sup_y d_t(y) + \rho(t) + b_1\pi(t), \quad \text{with } b_1 := \|v^*\|,$$

so that

$$\begin{aligned} d_{t+1}^+ &\leq d_t^+ + \rho(t) + b_1\pi(t) \\ &\leq d_0^+ + \sum_{i=0}^{t}[\rho(i) + b_1\pi(i)] \\ &\leq d_0^+ + \overline{\rho} + b_1\overline{\pi} \quad \text{for all } t, \end{aligned}$$

which implies that the functions d_t are uniformly bounded from above.

A similar argument yields that the d_t are uniformly bounded from below, since

$$\begin{aligned} d_{t+1}^- &\geq d_t^- - \rho(t) - b_1\pi(t) \\ &\geq d_0^- - (\overline{\rho} + b_1\overline{\pi}) \quad \text{for all } t, \end{aligned}$$

and the desired result follows. □

Note that, from the proof of Lemma 6.4, $\text{sp}(d_{t+1}) \leq \text{sp}(d_t) \leq \text{sp}(d_0)$ for all t. We will prove below that $\text{sp}(d_t)$ converges to zero.

In the main result of this section, Theorem 6.6 below, we will use the following functions.

6.5 Definition. Let M be the constant in Lemma 6.4 and, for every $t \geq 0$, define

$$\gamma(t) := \rho(t) + M\pi(t), \quad \text{and} \quad \gamma^c(t) := \rho^c(t) + M\pi^c(t),$$

where

$$\rho^c(t) := \sum_{i=t}^{\infty}\rho(i) \quad \text{and} \quad \pi^c(t) := \sum_{i=t}^{\infty}\pi(i).$$

The following theorem is the NVI-analogue of the value iteration results in Section 3.4.

6.6 Theorem. *Suppose that Assumptions* 6.1 *hold. Then:*

(a) *There exists a constant d such that*

$$\sup_x |d_t(x) - d| \leq (1+2M)D_t \quad \textit{for all } t \geq 0,$$

where

$$D_t := \max\{\rho^c([t/2]), \pi^c([t/2]), \alpha^{[t/2]}\}.$$

(b) $\sup_x |\overline{v}_t(x) - \overline{v}_{t-1}(x) - j^*| = \sup_x |d_t(x) - d_{t-1}(x)| \leq (1+2M)(D_t + D_{t-1})$ *for all* $t \geq 1$.

(c) $\sup_{x,y} |[\overline{v}_t(x) - \overline{v}_t(y)] - [v^*(x) - v^*(y)]| \leq \sup_{x,y} |d_t(x) - d_t(y)| \leq 2(1+2M)D_t$ *for all* $t \geq 0$, *where the* sup *is over all x and y in X.*

(d) *For each* $t \geq 0$, *let* $\overline{f}_t \in \mathbf{F}$ *be such that* $\overline{f}_t(x) \in A(x)$ *maximizes the r.h.s. of* 6.2 *for every* $x \in X$, *and let* $\overline{\delta} := \{\overline{f}_t\}$. *Then* $\overline{\delta}$ *is an average optimal Markov policy for the limiting MCM* (X, A, q, r).

Note that, when the policy $\overline{\delta}$ is used, the action at time t is

$$a_t = \overline{f}_t(x_t).$$

We shall call $\overline{\delta}$ an *average-reward NVI policy.* (NVI policies for *discounted* reward problems were introduced in Chapter 2.)

Proof. (a) Let c_t be the functions defined in 6.3; then a direct computation using the definition of the operators T and T_t yields

$$\begin{aligned} T_t c_t(x) - Tc_t(x) &\leq \rho(t) + M\pi(t) \\ &= \gamma(t) \qquad \text{by (6.5), for all } t \geq 0 \text{ and } x \in X. \end{aligned} \tag{1}$$

Therefore,

$$\begin{aligned} c_{t+1}(x) &= \overline{v}_{t+1}(x) - (t+1)j^* \\ &= T_t c_t(x) - j^* \\ &= Tc_t(x) - j^* + T_t c_t(x) - Tc_t(x) \\ &\leq Tc_t(x) - j^* + \gamma(t) \quad \text{for all } t \geq 0 \text{ and } x \in X, \end{aligned}$$

so that, for all $n \geq 1$,

$$\begin{aligned} c_{t+n}(x) &\leq T^n c_t(x) - nj^* + \sum_{i=t}^{t+n-1} \gamma(i) \\ &\leq T^n c_t(x) - nj^* + \gamma^c(t). \end{aligned}$$

Subtracting $v^*(x)$ on each side of this inequality, Definition 6.3 yields

$$d_{t+n}(x) \leq e_{t,n}(x) + \gamma^c(t) \quad \text{for all } t \geq 0, n \geq 1, \text{and } x \in X,$$

where

$$\begin{aligned} e_{t,n}(x) &:= T^n c_t(x) - T^n v^*(x) \quad \text{(cf. 4.4)} \\ &= T^n c_t(x) - nj^* - v^*(x). \end{aligned}$$

Similarly, if instead of inequality (1) we use

$$T_t c_t(x) - T c_t(x) \geq -\gamma(t),$$

we then obtain

$$d_{t+n}(x) \geq e_{t,n}(x) - \gamma^c(t).$$

Therefore,

$$\begin{aligned} |d_{t+n}(x) - e_{t,n}(x)| &\leq \gamma^c(t) \\ &= \rho^c(t) + M\pi^c(t) \quad \text{(by 6.5)}. \end{aligned}$$

On the other hand, the span-contraction property of T yields that (as in Lemma 4.6)

$$\begin{aligned} \text{sp}(e_{t,n}) &\leq \alpha^n \cdot \text{sp}(c_t - v^*) \\ &= \alpha^n \cdot \text{sp}(e_{t,0}), \end{aligned}$$

and there exists a constant d such that, for every $t \geq 0$ and $n \geq 1$,

$$\begin{aligned} \sup_x |e_{t,n}(x) - d| &\leq \alpha^n \cdot \text{sp}(e_{t,0}) \\ &\leq \alpha^n \cdot \text{sp}(d_t) \\ &\leq M\alpha^n. \end{aligned}$$

Thus

$$\begin{aligned} |d_{t+n}(x) - d| &\leq |d_{t+n}(x) - e_{t,n}(x)| + |e_{t,n}(x) - d| \\ &\leq \rho^c(t) + M\pi^c(t) + M\alpha^n \\ &\leq (1+2M) \cdot \max\{\rho^c(t), \pi^c(t), \alpha^n\}. \end{aligned}$$

Now let $m = t + n$, with $t = [m/2]$ and $n = m - t \geq [m/2]$; then the inequality becomes

$$\begin{aligned} |d_m(x) - d| &\leq (1+2M) \cdot \max\{\rho^c([m/2]), \pi^c([m/2]), \alpha^{[m/2]}\} \\ &= (1+2M) \cdot D_m, \end{aligned}$$

which proves part (a).

(b) and (c): Both follow from Definition 6.3 and part (a).

(d) As usual, by Theorem 2.2(d), it suffices to verify that, as $t \to \infty$,

$$\sup_x |\phi(x, \overline{f}_t(x))| \to 0. \tag{2}$$

To prove this, let us simplify the notation writing $\overline{f}_t(x)$ as a_t. Then by the definition of ϕ in Theorem 2.2(c), we obtain

$$\phi(x, a_t) = r(x, a_t) + \int [v^*(y) - v^*(x)]\, q(dy \mid x, a) - j^*, \tag{3}$$

whereas 6.2 and the definition of $\overline{f}_t(x) = a_t$ yield

$$\overline{v}_{t+1}(x) - \overline{v}_t(x) = r_t(x, a_t) + \int [\overline{v}_t(y) - \overline{v}_t(x)]\, q(dy \mid x, a).$$

Finally, on the r.h.s. of (3), add and subtract $\overline{v}_{t+1}(x) - \overline{v}_t(x)$ to obtain

$$|\phi(x, a_t)| \le \rho(t) + 2\|v^*\|\pi(t) + |\overline{v}_{t+1}(x) - \overline{v}_t(x) - j^*|$$
$$+ \sup_y |[\overline{v}_t(y) - \overline{v}_t(x)] - [v^*(y) - v^*(x)]|,$$

so that (2) follows from parts (b) and (c) and the assumptions. This completes the proof of the theorem. □

Note that the results in Section 3.5 are considerably sharper than Lemma 6.4 and Theorem 6.6. But, on the other hand, the NVI scheme 6.2 has the advantage of being recursive and, moreover, we can specialize it to obtain other approximating schemes, such as the following.

Nonstationary Successive Averagings

A nonstationary version of the "successive averagings" functions in 4.9 and 4.10 is obtained as follows.

Let $\overline{u}_0 := 0$, and let $\overline{u}_t := t^{-1}\overline{v}_t$ for $t \ge 1$. From the NVI equation 6.2, the functions $\overline{u}_t$ satisfy

6.7 $\qquad \overline{u}_{t+1}(x) = \overline{Q}_t \overline{u}_t(x)$ for all $t \ge 0$ and $x \in X$,

where $\overline{Q}_t$ is the operator on $B(X)$ defined by

$$\overline{Q}_t v(x) := \max_{a \in A(x)} \left\{ (t+1)^{-1} r_t(x, a) + t(t+1)^{-1} \int v(y)\, q_t(dy \mid x, a) \right\}.$$

Clearly, for each $t \ge 0$, the operator $\overline{Q}_t$ is a contraction on $B(X)$ with modulus $t/(t+1)$, and therefore, there exists a unique (fixed point) function $\overline{u}_t^* \in B(X)$ such that

6.8 $\qquad \overline{u}_t^* = \overline{Q}_t \overline{u}_t^*$ for all t;

and also, as a consequence of Theorem 6.6 we can obtain a nonstationary version of Corollary 4.11; namely:

6.9 Corollary. *Suppose that the assumptions of Theorem 6.6 hold. Then, as $t \to \infty$, each of the following sequences converges to zero:*

(a) $\sup_x |\overline{u}_t(x) - j^*|$

(b) $\|\overline{u}_t - \overline{u}_t^*\|$

(c) $\sup_x |\overline{u}_t^*(x) - j^*|$.

Furthermore, with the obvious changes in notation, Remark 4.12 also holds in the present, nonstationary case.

Discounted-Like NVI

A review of the results above would show that the measures $q_t(\cdot \mid k)$ may be sub-probabilities, that is, $q_t(X \mid k) \leq 1$, provided they satisfy the requirements in Assumption 6.1. In particular, we may take

$$q_t'(\cdot \mid k) := \beta_t q(\cdot \mid k) \quad \text{for all } t \geq 0 \text{ and } k \in \mathbf{K},$$

where $\{\beta_t\}$ is an increasing sequence of positive numbers converging to 1, in which case, the corresponding condition in 6.1 becomes

$$\overline{\pi}' := \sum (1 - \beta_t) < \infty.$$

In such a case, the NVI functions, which we now denote by h_t—instead of $\overline{v}_t$ as in 6.2—are defined by

6.10 $$h_{t+1}(x) := U_t h_t(x) \quad \text{for } t \geq 0,$$

where $h_0 \in B(X)$ is arbitrary and U_t is the operator on $B(X)$ defined as

$$U_t v(x) := \max_{a \in A(x)} \left\{ r_t(x,a) + \beta_t \int v(y)\, q(dy \mid x,a) \right\}.$$

Thus the U_t are like the discounted-reward operators T_t in Section 2.4, except that now the discount rate β_t is "time-dependent." In particular, U_t is a contraction operator with modulus β_t, and therefore, for each $t \geq 0$, there exists a unique (fixed point) $h_t^* \in B(X)$ such that

$$h_t^* = U_t h_t^*,$$

and then Lemma 6.4, Theorem 6.6 and Corollary 6.9 can be re-stated, with appropriate changes, in terms of the functions h_t and h_t^*. We leave the details to the reader.

Alternatively, we may take

6.11 $$q_t''(\cdot \mid k) := \beta_t q_t(\cdot \mid k),$$

where β_t and $q_t(\cdot \mid k)$ as above, and

$$\pi''(t) := \sup_k \|q_t''(\cdot \mid k) - q(\cdot \mid k)\| \leq \beta_t \pi(t) + (1 - \beta_t).$$

In this case, the corresponding condition in Assumption 6.1 is that (since $\beta_t < 1$)

$$\overline{\pi}'' := \overline{\pi} + \overline{\pi}' < \infty.$$

Several authors have used discounted-NVI functions. For instance, Hordijk and Tijms (1975) study 6.10 for MCM's with *finite* state and control spaces, and $r_t(k) = r(k)$ independent of t. On the other hand, Gordienko (1985) introduces an adaptive policy for discrete-time systems of the form

$$x_{t+1} = F(x_t, a_t, \xi_t),$$

where the i.i.d. disturbance sequence $\{\xi_t\}$ has unknown distribution, similar to what we did in Section 2.6 for discounted-reward problems. In Gordienko's paper, $r_t(k) = r(k)$ for all t, and $q_t(\cdot \mid k)$ is the empirical distribution θ_t defined in Section 2.6, equation 6.9; his assumptions are similar to those in Section 2.6 plus an ergodicity condition of the form 3.1(5) in the present chapter.

We will now use the results in this section and in Section 3.5 to study adaptive MCM's.

3.7 Adaptive Control Models

We now consider MCM's $(X, A, q(\theta), r(\theta))$ in which the transition law and the one-step reward function, say $q(\cdot \mid k, \theta)$ and $r(k, \theta)$, depend on a parameter θ that takes values in a Borel space Θ.

For each (fixed) $\theta \in \Theta$, everything remains essentially the same as in Sections 3.1 to 3.4 except for notational changes. For instance, instead of the average reward $J(\delta, x)$ in 1.1, we now have

$$J(\delta, x, \theta) := \liminf_{n\to\infty} n^{-1} E_x^{\delta,\theta} \sum_{t=0}^{n-1} r(x_t, a_t, \theta),$$

where $E_x^{\delta,\theta}$ denotes the expectation with respect to the probability measure $P_x^{\delta,\theta}$; see Section 1.1.

The program for this section is as follows. Firstly, we summarize some facts from Sections 3.1 to 3.4; the idea is to put the parametric MCM's in the appropriate setting. Secondly, we re-state some of the approximation results in Sections 3.5 and 3.6, and finally, those results are used to introduce several average-optimal *adaptive* policies, i.e., policies for MCM's depending on unknown parameters.

Preliminaries

To begin with, we suppose valid the θ-analogue of Assumptions 5.1, which include Assumptions 2.1 [cf. 7.1(a), (b) and (c) below], the ergodicity condition 3.1(4) [cf. 7.1(d)], and Assumption 5.1(c) [cf. 7.1(e)].

7.1 Assumptions.

(a) Same as 2.1(a).

(b) $r(k,\theta)$ is a measurable function on $\mathbf{K}\Theta$ such that $|r(k,\theta)| \leq R < \infty$ for all $k = (x,a) \in \mathbf{K}$ and $\theta \in \Theta$, and, moreover, $r(x,a,\theta)$ is a continuous function of $a \in A(x)$ for every $x \in X$ and $\theta \in \Theta$.

(c) $q(\cdot \mid k,\theta)$ is a stochastic kernel on X given $\mathbf{K}\Theta$ such that

$$\int_X v(y,\theta)\, q(dy \mid x,a,\theta)$$

is a continuous function of $a \in A(x)$ for every $x \in X$, $\theta \in \Theta$, and $v \in B(X\Theta)$.

(d) $\sup_{\theta,k,k'} \|q(\cdot \mid k,\theta) - q(\cdot \mid k',\theta)\| \leq 2\alpha$, with $\alpha < 1$, where the sup is over all k and k' in $\mathbf{K}$ and $\theta \in \Theta$.

(e) For any $\theta \in \Theta$ and any sequence $\theta_t \in \Theta$ that converges to θ, the sequences $\rho(t,\theta)$ and $\pi(t,\theta)$ converge to zero, where

$$\rho(t,\theta) := \sup_k |r(k,\theta_t) - r(k,\theta)|$$

and

$$\pi(t,\theta) := \sup_k \|q(\cdot \mid k,\theta_t) - q(\cdot \mid k,\theta)\|.$$

From Assumptions 7.1(a)–(d) and Corollary 3.6, there exist bounded functions $j^*(\theta) \in B(\Theta)$ and $v^*(x,\theta) \in B(X\Theta)$ which are a solution to the θ-optimality equation (OE)

7.2 $$j^*(\theta) + v^*(x,\theta) = T_\theta v^*(x,\theta) \quad \text{for all } x \in X,$$

where, for each $\theta \in \Theta$, T_θ is the operator on $B(X)$ defined, for v in either $B(X)$ or $B(X\Theta)$, by

7.3 $$T_\theta v(x) := \max_{a \in A(x)} \left\{ r(x,a,\theta) + \int_X v(y)\, q(dy \mid x,a,\theta) \right\}.$$

This in turn implies that, for each $\theta \in \Theta$, the conclusions (a)–(d) in Theorem 2.2 hold, and on the other hand, making the substitutions

7.4 $$r_t(k) := r(k,\theta_t) \quad \text{and} \quad q_t(\cdot \mid k) := q(\cdot \mid k,\theta_t),$$

$$j_t^* := j^*(\theta_t),\ v_t^*(x) := v^*(x,\theta_t),\ \text{and}\ f_t^*(x) := f^*(x,\theta_t),$$

the θ-version of Theorem 5.7 yields the following.

The Principle of Estimation and Control (PEC)

Let $f^*(x,\theta)$ be a measurable function from $X\Theta$ to A such that, for each $\theta \in \Theta$ and $x \in X$, the action $f^*(x,\theta) \in A(x)$ maximizes the r.h.s. of the θ-(OE) 7.2. By Theorem 2.2(b), $f^*(\cdot,\theta) \in \mathbf{F}$ is an optimal stationary policy for the θ-MCM, and then using the substitution 7.4 above and Theorem 5.7 we conclude:

7.5 Corollary. *Suppose that Assumptions* 7.1(a)–(e) *hold and let* $\{\theta_t\}$ *be any sequence converging to* θ *in* Θ. *Then*

(a) $\lim_{t\to\infty} |j^*(\theta_t) - j^*(\theta)| = 0$.

(b) $\text{sp}(v^*(\cdot,\theta_t) - v^*(\cdot,\theta)) \leq b_0(\theta) \cdot \max\{\rho(t,\theta), \pi(t,\theta)\}$, *where* $b_0(\theta) := 2(1 + \|v^*(\cdot,\theta)\|)/(1-\alpha)$.

(c) *The policy* $\{f^*(\cdot,\theta_t)\}$ *that takes the action*

$$a_t := f^*(x_t,\theta_t) \quad \textit{at time } t$$

is average-optimal for the θ-*MCM* $(X, A, q(\theta), r(\theta))$.

(d) *Let* $\{\hat{\theta}_t\}$ *be a sequence of strongly consistent* (*SC*) *estimators of* θ (*see Definition* 5.14 *in Chapter* 2), *and let* $\delta_\theta^* = \{\delta_t^*\}$ *be the policy defined by*

$$\delta_t^*(h_t) := f^*(x_t, \hat{\theta}_t(h_t)) \quad \textit{for every } h_t \in H_t \textit{ and } t \geq 0,$$

where H_t *is the set of t-histories. Then, from part* (c), δ_θ^* *is average-optimal for the* θ-*MCM.*

We call δ_θ^* a PEC *adaptive policy;* for a further discussion on it, see Section 2.5 (in particular, the Remarks 5.11 on the construction of the PEC policy).

Nonstationary Value Iteration (NVI)

We now require the θ-analogue of Assumptions 6.1. Thus, *in addition to* 7.1, let us suppose:

7.6 Assumption. For any θ and θ_t as in 7.1(e),

$$\overline{\rho}(\theta) := \sum_t \rho(t,\theta) < \infty \quad \text{and} \quad \overline{\pi}(\theta) := \sum_t \pi(t,\theta) < \infty.$$

And then, instead of the NVI functions $\overline{v}_t(x)$ in 6.2, define $\overline{v}_0(\cdot) := 0$ and

$$\overline{v}_{t+1}(x,\theta_t) := T_{\theta_t}\overline{v}_t(x,\theta_{t-1}) \quad \text{for all } t \geq 0.$$

That is,

$$\overline{v}_1(x,\theta_0) := \max_{a\in A(x)} r(x,a,\theta_0),$$

and for $t \geq 1$,

$$\text{7.7} \qquad \overline{v}_{t+1}(x,\theta_t) := \max_{a \in A(x)} \left\{ r(x,a,\theta_t) + \int \overline{v}_t(y,\theta_{t-1})\, q(dy \mid x,\theta_t) \right\}.$$

Thus, the θ-analogue of the NVI policy $\overline{\delta} = \{\overline{f}_t\}$ in Theorem 6.6(d) can be defined as follows.

For each $t \geq 0$, let $\overline{f}_t(\cdot,\theta_t) \in \mathbf{F}$ be such that $\overline{f}_t(x,\theta_t) \in A(x)$ maximizes the r.h.s. of 7.7 for every $x \in X$; note that

$$\overline{f}_0(x,\theta_0) = \arg \max_{a \in A(x)} r(x,a,\theta_0).$$

Strictly speaking we should write $\overline{f}_t(x,\theta_t)$ as $\overline{f}_t(x,\theta_t,\theta_{t-1},\ldots,\theta_0)$, but we shall keep the shorter notation $\overline{f}_t(x,\theta_t)$.

Using the substitutions 7.4 again, we can rewrite Theorem 6.6 in the following form.

7.8 Corollary. *Suppose that Assumptions* 7.1 *and* 7.6 *hold, and let* $\{\theta_t\}$ *be any sequence converging to* $\theta \in \Theta$. *Then:*

(a) *There exists a constant* $d(\theta)$ *such that*

$$\sup_x |d_t(x,\theta) - d(\theta)| \to 0 \quad \text{as} \quad t \to \infty,$$

where (cf. Definition 6.3) $d_t(x,\theta) := \overline{v}_t(x,\theta_{t-1}) - tj^*(\theta) - v^*(x,\theta)$.

(b) $\sup_x |\overline{v}_{t+1}(x,\theta_t) - \overline{v}_t(x,\theta_{t-1}) - j^*(\theta)| \to 0$ *as* $t \to \infty$.

(c) $\sup_{x,y} |[\overline{v}_{t+1}(x,\theta_t) - \overline{v}_{t+1}(y,\theta_t)] - [v^*(x,\theta) - v^*(y,\theta)]| \to 0$ *as* $t \to \infty$.

(d) *Let* $\{\hat{\theta}_t\}$ *be a sequence of strongly consistent estimators of* θ *(Chapter* 2, *Definition* 5.14), *and let* $\overline{\delta}_\theta = \{\overline{\delta}_t\}$ *be the policy defined by*

$$\overline{\delta}_t(h_t) := \overline{f}_t(x_t,\hat{\theta}_t(h_t)) \quad \text{for every} \quad h_t \in H_t \quad \text{and} \quad t \geq 0.$$

Then [by Theorem 6.6(d)] $\overline{\delta}_\theta$ *is an average-optimal policy for the* θ-*MCM.*

In Corollary 7.8(a), (b) and (c), convergence rates, such as $(1+2M)D_t$ in Theorem 6.6, can be computed in terms of the sequences $\rho(t,\theta)$ and $\pi(t,\theta)$ in Assumptions 7.1(e) and 7.6.

We call $\overline{\delta}_\theta$ an NVI *adaptive policy.* Other NVI adaptive policies can be defined via suitable variants of the NVI scheme in Section 6: for instance, Baranov (1981) uses the nonstationary successive averagings 6.7 and 6.8, and Gordienko (1985) uses the discounted-NVI defined in terms of 6.11. For adaptive policies in terms of the value-iteration functions v_t in Section 3.4, see Acosta Abreu (1987a).

The PEC adaptive policy in Corollary 7.5(d) has been studied by many authors: Kurano (1972), Mandl (1974), Georgin (1978b), and Kolonko (1982a), among others.

3.8 Comments and References

We presented in this chapter some results on approximations and adaptive policies for MCM's with the average reward criterion. The chapter is essentially an expanded (and slightly improved) version of Hernández-Lerma (1987), but other specific references were provided in each section. We now mention some other related works.

The dynamic programming (DP) or optimality equation (OE) in Theorem 2.2 was originally studied by Bellman (1957, 1961) for some special cases, but for general (Borel) state space, parts (a) and (b) in that theorem are due to Ross (1968). Ross (1970, 1971, 1983) also introduced several "pathological" examples for which average-optimal policies do not exist, and examples for which no stationary policy is optimal. Parts (c) and (d) in Theorem 2.2 are probably due to Mandl (1974) for finite-state MCM's, and to Georgin (1978a) for the Borel case.

Other important contributions to the Borel case—including the use of ergodicity conditions such as those in Section 3.3—are made by Gubenko and Statland (1975), Kurano (1986), Rieder (1979), Tijms (1975), among others. All of these authors have used in one way or other the fact that the ergodicity conditions 3.1 are closely related to some contraction property of the DP operator, as in Lemma 3.5 and the Comments 3.7. For discussions on the relationships between the ergodicity conditions 3.1 and Döeblin's condition, see, e.g., Doob (1953), Tijms (1975), Ueno (1957). Hernández-Lerma and Doukhan (1988) give conditions under which a system of the form

$$x_{t+1} = f(x_t, a_t) + \xi_t$$

satisfies 3.1(4). Other applications of ergodicity or "contraction" properties include the determination of forecast horizons and the analysis of discounted-reward problems (see Chapter 2) with discount factor greater than 1; see, e.g., Hernández-Lerma and Lasserre (1988), Hinderer and Hübner (1977), and Hopp et al. (1986). Hernández-Lerma, Esparza and Duran (1988) use 3.1(4) to obtain consistent recursive estimators of the transition probability density of non-controlled Markov processes. Finally, Cavazos-Cadena (1988 a,b,c) considers denumerable-state MCM's and shows, among other important results, that some ergodicity conditions are *necessary* for the existence of a bounded solution to the (OE).

The successive approximations or value iteration approach in Section 3.4 has a long list of contributors going back to Bellman (1957, 1961), and including D.J. White (1963), Federgruen and Tijms (1978), Hordijk et al. (1975), Schweitzer (1971), Tijms (1975), to name a few.

References to Sections 3.5 and 3.6 on approximation of MCM's were given in the text; for discounted reward problems, see Section 2.4. In Chapter 6 we study the approximation of MCM's by discretizations of the state space.

At the end of Section 3.7 (and also in Section 2.7) we cited related works on adaptive MCM's, in particular, the pioneering works of Kurano (1972) and Mandl (1974). There are many other approaches for adaptive average-reward problems, such as those by Borkar and Varaiya (1982), and El-Fattah (1981). For an extensive survey of pre-1984 works see Kumar (1985). For *nonparametric average*-reward problems, see Gordienko (1985), or Hernández-Lerma and Duran (1988); Kurano (1983, 1987), and White and Eldeib (1987) study *finite*-state MCM's with uknown transition *matrix*, which can be considered as being "parametric" problems. For references on the nonparametric *discounted* case, see Sections 2.6 and 2.7.

Note added in proof. In the paper by Cavazos-Cadena and Hernández-Lerma (1989) it has been proved that the optimality of the NVI adaptive policy $\overline{\delta}_\theta$ in Corollary 7.8(d) can be obtained *without* Assumption 7.6; that is, Assumptions 7.1 are sufficient.

4

Partially Observable Control Models

4.1 Introduction

Stochastic control systems with partial state information appear in many important problems in engineering, economics, population processes, learning theory and many other areas. The key feature of these systems is that, in contrast with the situation studied in previous chapters, the state x_t of the system cannot be observed directly; instead, we get information about x_t through an observation or measurement process y_t. A typical model is of the form

1.1 (a) $x_{t+1} = F(x_t, a_t, \xi_t), \quad t = 0, 1, \ldots,$

(b) $y_t = G(a_{t-1}, x_t, \eta_t), \quad t = 1, 2, \ldots,$

(c) $y_0 = G_0(x_0, \eta_0),$

where x_t, a_t and y_t are, respectively, the state, the control and the observation at time t; $\{\xi_t\}$ is the so-called state-disturbance process, and $\{\eta_t\}$ is the observation (or measurement) noise. However, in some applications—as in learning theory or machine replacement/quality control problems—it is more convenient to consider a Partially Observable (Markov) Control Model (or PO-CM for short) as defined in Section 4.2 below.

The main objective in this chapter is to extend the adaptive control results in previous chapters to PO-CM's depending on unknown parameters. We will do this as follows.

Summary

In Section 4.2, we begin with the definition of a PO-CM, and then we introduce the partially observable (PO) control problem. An important difference between the PO control problem and the standard "completely observable" (CO) problem in previous chapters is that the policies for the former case are defined in terms of the "observable" histories (say, $y_0, a_0, y_1, a_1, \ldots$), and not in terms of the (unobservable) state process $\{x_t\}$.

In Section 4.3, we transform the PO control problem into a CO control problem, in which the new "state" process z_t is the conditional (or *a posteriori*) distribution of the unobservable state x_t, given the observable history up to time t. The two problems turn out to be equivalent (Proposition 3.13).

This equivalence is used in Section 4.4 to obtain optimality conditions for the PO control problem in terms of the new CO problem (Theorem 4.3).

In Section 4.5 we consider PO-CM's depending on unknown parameters. The results in Sections 4.3 and 4.4 are combined with those in Chapter 2 to obtain adaptive policies for PO-CM(θ), where θ denotes the unknown parameter.

We conclude in Section 4.6 with some comments on possible extensions and related matters.

In the CO control problem of Section 4.3, the state space is $\mathbf{Z} := \mathbf{P}(X)$, the space of probability measures on X with the topology of weak convergence, where X is the state space in the original PO-CM. For the reader's convenience, some properties of $\mathbf{P}(X)$ are summarized in Appendix B. (We will use, in particular, the fact that $\mathbf{P}(X)$ is a Borel space whenever X is a Borel space.)

4.2 PO-CM: Case of Known Parameters

As already noted, 1.1 represents a typical PO system. In more generality we have the following.

2.1 Definition. A partially observable control model (PO-CM) is specified by $(X, Y, A, P, Q, Q_0, p, r)$, where:

(a) X, the *state space,* is a Borel space.

(b) Y, the *observation set,* is a Borel space.

(c) A, the *control* (or *action*) *set,* is a Borel space.

(d) $P(dx' \mid x, a)$, the *state transition law,* is a stochastic kernel on X given XA.

(e) $Q(dy \mid a, x)$, the *observation kernel,* is a stochastic kernel on Y given AX.

(f) $Q_0(dy \mid x)$, the *initial observation kernel,* is a stochastic kernel on Y given X.

(g) $p \in \mathbf{P}(X)$ is the (*a priori*) *initial distribution.*

(h) $r \in B(XA)$ is the one-step reward function.

The PO-CM evolves as follows. At time $t = 0$, the initial (unobservable) state x_0 has a given *a priori* distribution p, and the initial observation y_0

is generated according to the initial observation kernel $Q_0(\cdot \mid x_0)$. If at time t the state of the system is x_t and the control $a_t \in A$ is applied, then a reward $r(x_t, a_t)$ is received and the system moves to state x_{t+1} according to the transition law $P(dx_{t+1} \mid x_t, a_t)$; the observation y_{t+1} is generated by the observation kernel $Q(dy_{t+1} \mid a_t, x_{t+1})$.

To illustrate these ideas, consider the system 1.1, where x_t, y_t and a_t take their respective values in Borel spaces, X, Y and A. Assume $\{\xi_t\}$ and $\{\eta_t\}$ are sequences of mutually independent i.i.d. random elements with values in Borel spaces S and N, and distributions $\mu \in \mathbf{P}(S)$ and $\nu \in \mathbf{P}(N)$, respectively; F, G and G_0 are given measurable functions, and x_0 is independent of $\{\xi_t\}$ and $\{\eta_t\}$. Then the state transition law is given by

2.2
$$P(B \mid x, a) = \int_S 1_B[F(x, a, s,)]\, \mu(ds)$$

for every Borel set B in X, and $x \in X$ and $a \in A$. Similarly, if $a_t = a$ and $x_{t+1} = x$, the observation kernel is given by

2.3
$$Q(C \mid a, x) = \int_N 1_C[G(a, x, n)]\, \nu(dn) \quad \text{for all} \quad C \in \mathcal{B}(Y),$$

whereas, if $x_0 = x$,

$$Q_0(C \mid x) = \int_N 1_C[G_0(a, x, n)]\, \nu(dn) \quad \text{for all} \quad C \in \mathcal{B}(Y).$$

Thus a realization of the partially observable (PO) system looks like

$$(x_0, y_0, a_0, x_1, y_1, a_1, \ldots) \in \Omega := (XYA)^\infty,$$

where x_0 has a given distribution $p \in \mathbf{P}(X)$, and $\{a_t\}$ is a control sequence in A determined by a control policy.

Now to define a policy we cannot use the (unobservable) states $x_0, x_1, \ldots$. We thus introduce the *observable histories* $h_0 := (p, y_0) \in H_0$ and

2.4
$$h_t := (p, y_0, a_0, \ldots, y_{t-1}, a_{t-1}, y_t) \in H_t \quad \text{for all} \quad t \geq 1,$$

where $H_0 := \mathbf{Z}Y$ and $H_t := H_{t-1}AY$ if $t \geq 1$, with $\mathbf{Z} := \mathbf{P}(X)$. Then a *policy* for the PO-CM in Definition 2.1 is defined as a sequence $\pi = \{\pi_t\}$ such that, for each t, π_t is a stochastic kernel on A given H_t. The set of all policies is denoted by Π.

A policy $\pi \in \Pi$ and an initial distribution $p_0 = p \in \mathbf{Z}$, together with the stochastic kernels P, Q and Q_0 in Definition 2.1, determine on the space Ω of all possible realizations of the PO system a probability measure P_p^π given by

$$P_p^\pi(dx_0 dy_0 da_0 dx_1 dy_1 da_1 \ldots) = p(dx_0) Q_0(dy_0 \mid x_0) \pi_0(da_0 \mid p, y_0)$$
$$\cdot\, P(dx_1 \mid x_0, a_0) Q(dy_1 \mid a_0, x_1) \pi_1(da_1 \mid p, y_0, a_0, y_1) \ldots,$$

or, using the abbreviated notation for product measures in Proposition C.3, Appendix C,

2.5 $$P_p^\pi = pQ_0\pi_0PQ\pi_1PQ\pi_2\ldots.$$

The expectation with respect to this probability measure is denoted by E_p^π.

The PO Control Problem

To define the control problem we need to specify a performance criterion. Here we shall restrict ourselves to discounted reward problems, but other criteria can be treated similarly (references are given at the end of the chapter).

Let

2.6 $$J(\pi,p) := E_p^\pi \sum_{t=0}^\infty \beta^t r(x_t,a_t)$$

be the expected total discounted reward when policy $\pi \in \Pi$ is used and the initial distribution is $p \in \mathbf{Z}$. [Recall that $\mathbf{Z} := \mathbf{P}(X)$.] As usual, the number $\beta \in (0,1)$ is the discount factor. The PO control problem is then to find a policy $\pi^* \in \Pi$ such that

$$J(\pi^*,p) = J^*(p) \quad \text{for all} \quad p \in \mathbf{Z},$$

where

$$J^*(p) := \sup_{\pi\in\Pi} J(\pi,p), \quad p \in \mathbf{Z},$$

is the *optimal reward function.*

4.3 Transformation into a CO Control Problem

Following the program described in the Introduction, we shall now transform the PO control problem in Section 4.2 into a new *completely observable* (CO) problem. Since this transformation is quite standard (references are given in Section 4.6) we will be brief and present only the main facts.

The idea is to introduce a new control model in which the state process, denoted by $\{z_t\}$, takes values in the space $\mathbf{Z} = \mathbf{P}(X)$ of probability measures on X and evolves according to an equation of the form

3.1 $$z_{t+1} = H(z_t,a_t,y_{t+1}), \text{ where } t = 0,1,\ldots, \quad \text{and} \quad z_0 = H_0(p,y_0).$$

In 3.1 $\{a_t\}$ and $\{y_t\}$ are the control and observation sequences in the original PO-CM, and $p \in \mathbf{Z}$ is the *a priori* distribution of the initial state x_0. To obtain H and H_0 we use the following general result on the decomposition (or "factorization") of a probability measure on a product space.

3.2 Lemma. *Let X, Y and W be Borel spaces and let $R(d(x,y)\,|\,w)$ be a stochastic kernel on XY given W. Then there exist stochastic kernels*

$H'(dx \mid w, y)$ and $R'(dy \mid w)$ on X given WY and on Y given W, respectively, such that

$$R(BC \mid w) = \int_C H'(B \mid w, y)\, R'(dy \mid w)$$

for all $B \in \mathcal{B}(X)$, $C \in \mathcal{B}(Y)$ and $w \in W$, where $R'(dy \mid w)$ is the marginal of $R(d(x,y) \mid w)$ on Y, i.e.,

$$R'(C \mid w) := R(XC \mid x), \quad C \in \mathcal{B}(Y).$$

Proof. Bertsekas and Shreve (1978), Corollary 7.27.1, p. 139; or Dynkin and Yushkevich (1979), p. 215; or Striebel (1975), Appendix A.1, etc.

Now, in Lemma 3.2, let X and Y be the state space and the observation set in the PO-CM (Definition 2.1), respectively, and take $W := \mathbf{Z}A$, where A is the control set and $\mathbf{Z} = \mathbf{P}(X)$ as above. Let $R(d(x,y) \mid z, a)$ be the stochastic kernel on XY given W such that, for all $B \in \mathcal{B}(X)$ and $C \in \mathcal{B}(Y)$,

3.3 $$R(BC \mid z, a) := \int_X \int_B Q(C \mid a, x')\, P(dx' \mid x, a)\, z(dx),$$

where P and Q are, respectively, the state transition law and the observation kernel in the PO-CM. Then, by the "decomposition" Lemma 3.2, there exists a stochastic kernel $H'(dx \mid z, a, y)$ on X given $WY = \mathbf{Z}AY$ such that, for all $B \in \mathcal{B}(X)$, $C \in \mathcal{B}(Y)$ and $(z, a) \in W$,

3.4 $$R(BC \mid z, a) = \int_C H'(B \mid z, a, y)\, R'(dy \mid z, a),$$

where $R'(C \mid z, a) := R(XC \mid z, a)$ is the marginal of $R(\cdot \mid z, a)$ on Y, that is,

3.5 $$R'(C \mid z, a) = \int_X \int_X Q(C \mid a, x')\, P(dx' \mid x, a)\, z(dx)$$

for all $C \in \mathcal{B}(Y)$ and $(z, a) \in W$.

It then follows, by the properties of stochastic kernels (Appendix C), that the function $H : \mathbf{Z}AY \to \mathbf{Z}$ defined by

3.6 $$H(z, a, y) := H'(\cdot \mid z, a, y)$$

is measurable, and therefore,

3.7 $$q(D \mid z, a) := \int_Y 1_D[H(z, a, y)]\, R'(dy \mid z, a),$$

where $D \in \mathcal{B}(\mathbf{Z})$ and $(z, a) \in W = \mathbf{Z}A$, defines a stochastic kernel on $\mathbf{Z}$ given $\mathbf{Z}A$.

A similar argument shows that if $Q_0(dy \mid x)$ is the initial observation kernel in the PO-CM (Definition 2.1), then there exists a stochastic kernel

$H'_0(dx \mid p, y)$ on X given $\mathbf{Z}Y$ such that, for all $B \in \mathcal{B}(X)$, $C \in \mathcal{B}(Y)$ and $p \in \mathbf{Z}$, the stochastic kernel

$$R_0(BC \mid p) := \int_B Q_0(C \mid x)\, p(dx)$$

on XY given $\mathbf{Z}$, can be decomposed as

3.8 $$R_0(BC \mid p) = \int_C H'_0(B \mid p, y)\, R'_0(dy \mid p),$$

where

$$R'_0(C \mid p) = \int_X Q_0(C \mid x)\, p(dx) \quad \text{for all } \; C \in \mathcal{B}(Y) \; \text{ and } \; p \in \mathbf{Z}.$$

Next, the function $H_0 : \mathbf{Z}Y \to \mathbf{Z}$ defined by

3.9 $$H_0(p, y) := H'_0(\cdot \mid p, y) \quad \text{for all } \; (p, y) \in \mathbf{Z}Y$$

is measurable, and therefore,

3.10 $$q_0(D \mid p) := \int_Y 1_D[H_0(p, y)]\, R'_0(dy \mid p), \text{ where } D \in \mathcal{B}(\mathbf{Z}) \text{ and } p \in \mathbf{Z}$$

defines a stochastic kernel on $\mathbf{Z}$ given $\mathbf{Z}$.

Remark. The functions H and H_0 in 3.1 are those defined by 3.6 and 3.9, respectively. In such a case, the random element z_t in the so-called *filtering equation* 3.1 can be interpreted as the *a posteriori* distribution of the unobservable state x_t given the observable history h_t. That is, for any policy π, any initial (*a priori*) distribution $p_0 = p \in \mathbf{Z}$, and any Borel subset B of X,

$$P_p^\pi(x_0 \in B \mid h_0) = H_0(h_0)(B) = z_0(B) \quad P_p^\pi\text{-a.s.},$$

where $h_0 = (p, y_0)$, and for all $t \geq 0$ and any observable history $h_{t+1} = (h_t, a_t, y_{t+1})$

$$P_p^\pi(x_{t+1} \in B \mid h_{t+1}) = H(z_t, a_t, y_{t+1})(B) = z_{t+1}(B) \quad P_p^\pi\text{-a.s.}$$

[See, e.g., Bertsekas and Shreve (1978), Section 10.3.1; Striebel (1975), Section 2.2; or Sawaragi and Yoshikawa (1970) when X and Y are denumerable.]

I-Policies

We will now define a new set of policies. To do this we note that, using the filtering equation 3.1, a sequence of observable histories $h_0 = (p, y_0), h_1, h_2, \ldots$, defines a process $z_0, z_1, \ldots$ with values in $\mathbf{Z}$, and therefore, a sequence of so-called *information vectors*

$$i_t := (z_0, a_0, \ldots, z_{t-1}, a_{t-1}, z_t) \in I_t, \quad t = 0, 1, \ldots,$$

where $I_t := \mathbf{Z}(A\mathbf{Z})^t$ for all $t = 0, 1, \ldots$, with $I_0 := \mathbf{Z}$. We then define an *information policy* (or *I-policy* for short) as a sequence $\delta = \{\delta_t\}$ such that, for each t, $\delta_t(da \mid i_t)$ is a stochastic kernel on A given I_t. We denote by Δ the set of all I-policies.

A sequence $\{f_t\}$ of measurable functions $f_t : \mathbf{Z} \to A$ is called a *Markov* I-policy and, as usual (cf. Section 1.2), we identify the set of all Markov I-policies with a subset of Δ. Also as usual, a Markov I-policy $\{f_t\}$ in which $f_t = f$ is independent of t is called a *stationary* I-policy and we refer to it simply as *the* stationary I-policy f.

We consider Δ as a subset of Π; that is, we consider an I-policy $\delta \in \Delta$ as a policy $\pi \in \Pi$. We can do this because any I-policy $\delta = \{\delta_t\}$ defines a policy $\pi^\delta = \{\pi_t^\delta\}$ in Π given by

$$\pi_t^\delta(\cdot \mid h_t) := \delta_t(\cdot \mid i_t(h_t)) \quad \text{for all } h_t \in H_t \text{ and } t \geq 0,$$

where $i_t(h_t) \in I_t$ is the information vector determined by the observable history h_t via 3.1. Thus δ and π^δ are equivalent in the sense that, for every $t \geq 0$, π_t^δ assigns the same conditional probability on A as that assigned by δ_t for any observable history h_t. It can also be shown the following [see, e.g., Sawaragi and Yoshikawa (1970), Rhenius (1974), Yushkevich (1976),...].

3.11 Proposition. Δ *is complete. That is, for any policy* $\pi \in \Pi$ *there exists an I-policy* $\delta \in \Delta$ *such that*

$$J(\delta, p) = J(\pi, p) \quad \textit{for all } p \in \mathbf{Z}.$$

The New Control Model

Given the PO-CM $(X, Y, A, P, Q, Q_0, p, r)$ in Definition 2.1, consider the completely observable (CO) control model $(\mathbf{Z}, A, q, q_0, r')$ with state space $\mathbf{Z} = \mathbf{P}(X)$, control set A as in the PO-CM, transition law $q(\cdot \mid z, a)$ in 3.7, initial distribution $q_0(\cdot \mid p)$ in 3.10 when the *a priori* distribution of the (unobservable) initial state x_0 is $p \in \mathbf{Z}$; and reward function $r' : \mathbf{Z}A \to \mathbf{R}$ defined by

$$r'(z, a) := \int_X r(x, a)\, z(dx).$$

The state process $\{z_t\}$ in this CO-CM is defined by 3.1.

The set of policies for the CO-CM is the set Δ of I-policies. An I-policy $\delta \in \Delta$ and an *a priori* distribution $p \in \mathbf{Z}$ for x_0 [together with the stochastic kernels q and $q_0(\cdot \mid p)$] define a probability measure

$$P_p'^{\delta} = q_0(\cdot \mid p)\, \delta_0 q \delta_1 q \delta_2 \ldots$$

on the space $\mathbf{Z}A\mathbf{Z}A\ldots$ of all sequence $(z_0, a_0, z_1, a_1, \ldots)$. The corresponding expectation is denoted by $E_p'^{\delta}$. Then the expected total discounted reward for the CO-CM is given by

$$V(\delta, p) := E_p'^{\delta} \sum_{t=0}^{\infty} \beta^t r'(z_t, a_t) \quad \text{for all } \delta \in \Delta \text{ and } p \in \mathbf{Z},$$

and the CO control problem is to determine an I-policy δ^* such that

$$V(\delta^*, p) = v^*(p) \quad \text{for all} \quad p \in \mathbf{Z},$$

where

3.12
$$v^*(p) := \sup_{\delta \in \Delta} V(\delta, p).$$

This control problem and the original PO problem are equivalent in the following sense (for a proof see any of the references for Proposition 3.11 above):

3.13 Proposition. *$V(\delta, p) = J(\delta, p)$ for all $\delta \in \Delta$ and $p \in \mathbf{Z}$.*

From Propositions 3.11 and 3.13 it follows that an I-policy is optimal (or Asymptotically Discount Optimal = ADO) for the CO control problem if, and only if, it is optimal (or ADO, respectively) for the original PO problem. In other words, results for the CO problem can be translated into results for the PO problem by replacing "policies" by "I-policies". We will follow this approach in the following section to obtain optimal and ADO policies for the PO control problem.

4.4 Optimal I-Policies

The main objective in this section is to give conditions under which the optimality results for discounted reward problems obtained in Sections 2.2 and 2.3 hold for the CO-CM $(\mathbf{Z}, A, q, q_0, r')$; these results are summarized in Theorem 4.3 below. To do this, we consider the PO-CM $(X, Y, A, P, Q, Q_0, p, r)$ in Definition 2.1 and assume that it satisfies the following.

4.1 Assumptions on PO-CM.

(a) A is compact.

(b) The one-step reward function $r(x, a) \in C(XA)$. [Recall that for any topological space W, $C(W)$ is the Banach space of real-valued, bounded, continuous functions on W with the supremum norm.]

(c) The state transition law $P(dx' \mid x, a)$ and the observation kernel $Q(dy \mid a, x)$ are *continuous* stochastic kernels (see Appendix C).

(d) The function $H(z, a, y)$ in 3.6 is continuous on $\mathbf{Z}AY$.

These assumptions, in particular 4.1(d), are briefly discussed in Remark 4.4 below. Now we note the following.

4.2 Lemma. *Assumptions* 4.1 *imply that the CO-CM* $(\mathbf{Z}, A, q, q_0, r')$ *satisfies:*

(a) *A is compact.*

(b) $r'(z, a) \in C(\mathbf{Z}A)$.

(c) *The stochastic kernel $q(dz' \mid z, a)$ is continuous.*

Proof. (a) This is the same as 4.1(a).

(b) This follows from 4.1(b) and Proposition C.2(b) in Appendix C, since, by definition,

$$r'(z, a) := \int r(x, a)\, z(dx).$$

(c) We want to show that for every $v \in C(\mathbf{Z})$, the function

$$v'(z, a) := \int_{\mathbf{Z}} v(z')\, q(dz' \mid z, a)$$

is continuous on $\mathbf{Z}A$. Now by Definition 3.7 of q, we can write

$$\begin{aligned} v'(z, a) &= \int_Y v[H(z, a, y)]\, R'(dy \mid z, a) \\ &= \int_X \int_X \int_Y v[H(z, a, y)] Q(dy \mid a, x')\, P(dx' \mid x, a)\, z(dx), \end{aligned}$$

and then by Assumptions 4.1(c) and (d), and repeated applications of Proposition C.2(b) in Appendix C, we obtain that v' is continuous. This completes the proof. □

Thus according to Lemma 4.2, the CO-CM (Z, A, q, q_0, r') satisfies the assumptions in Section 2.2 and therefore, substituting $B(X)$ by $C(\mathbf{Z})$, *all* the results in Sections 2.2 and 2.3 hold for CO-CM. In particular, in the present context, Theorems 2.2 and 2.8 in Chapter 2 can be re-stated as follows.

4.3 Theorem. *Suppose that Assumptions 4.1 hold. Then:*

(a) *The optimal reward function $v^* : \mathbf{Z} \to \mathbf{R}$ in 3.12 is the unique solution in $C(\mathbf{Z})$ of the dynamic programming equation (DPE)*

$$\begin{aligned} v^*(z) &= \max_{a \in A} \left\{ r'(z, a) + \beta \int_{\mathbf{Z}} v^*(z')\, q(dz' \mid z, a) \right\}, \quad z \in \mathbf{Z}, \\ &=: Tv^*(z), \end{aligned}$$

where T is the (contraction) operator on $C(\mathbf{Z})$ defined, for $v \in C(\mathbf{Z})$ and $z \in \mathbf{Z}$, by

$$Tv(z) := \max_{a \in A} \left\{ r'(z, a) + \beta \int v(z')\, q(dz' \mid z, a) \right\}.$$

(b) *A stationary I-policy f^* is optimal if and only if $f^*(z) \in A$ maximizes the r.h.s. of the DPE for all $z \in \mathbf{Z}$. The existence of one such policy f^* is insured under the present assumptions (by Lemma* 4.2 *and the Measurable Selection Theorem in Appendix* D, *Proposition* D.3).

The other results in Sections 2.2 and 2.3 can, of course, be translated into the present context. Instead of doing this, however, we will conclude this section with some remarks, and then consider PO-CM's with unknown parameters.

4.4 Remarks on Assumptions 4.1. Assumptions 4.1(a) and (b) are more or less standard in "discounted" dynamic programming (cf. Section 2.2), except that now we are considering the more restrictive setting of a control set $A(z) = A$ independent of the state, and reward function $r'(z,a)$ continuous in both variables, instead of only continuous in a and measurable in z, as in previous chapters. Both of these restrictions can be weakened [Rhenius (1974), Yushkevich (1976), Dynkin and Yushkevich (1979),...].

For specific PO-CM's such as 1.1, Assumption 4.1(c) can be expressed in terms of the system functions $F(x,a,s)$ and $G(a,x,n)$. For instance, under the conditions in the paragraph preceding equation 2.2, we can see that

4.5 continuity of $F(x,a,s)$ implies continuity of $P(dx' \mid x,a)$.

This follows from 2.2, since

$$\int_X h(x')\, P(dx' \mid x,a) = \int_S h[F(x,a,s)]\, \mu(ds)$$

is continuous in (x,a) for all $h \in C(X)$ if F is continuous. Similarly, using 2.3,

4.6 continuity of $G(a,x,n)$ implies continuity of $Q(dy \mid a,x)$.

Note that, in the general case, Assumption 4.1(c) implies that the stochastic kernel $R'(dy \mid z,a)$ in 3.5 is continuous; this is contained in the proof of Lemma 4.2(c).

Assumption 4.1(d) is on the CO-CM, in contrast with Assumptions 4.1(a), (b) and (c) which are conditions on the original PO-CM. Conditions sufficient for 4.1(d) can be easily derived for finite-dimensional systems of the form 1.1 with absolutely continuous disturbance distributions, as in Striebel (1975, Chapter 2). Another case is when the observation set Y is *denumerable* (with the discrete topology). In such a case, taking $C = \{y\}$ in 3.5 and writing

$$R'(\{y\} \mid z,a) = R'(y \mid z,a) \quad \text{and} \quad Q(\{y\} \mid a,x) = Q(y \mid a,x),$$

then from 3.3 and 3.4 we obtain

4.7 $$H'(B \mid z,a,y) = R'(y \mid z,a)^{-1} \int_X \int_B Q(y \mid a,x')P(dx' \mid x,a)z(dx)$$

for all Borel subset B of X, if $R'(y \mid z,a) \neq 0$; otherwise, we define $H'(\cdot \mid z,a,y)$ as an arbitrary probability measure in $\mathbf{Z} = \mathbf{P}(X)$. A direct calculation then shows that Assumption 4.1(c) implies 4.1(d). In other words, when the observation set Y is *denumerable,* Assumptions 4.1 require only parts (a), (b) and (c).

4.5 PO-CM's with Unknown Parameters

We now consider a PO-CM(θ) of the form

$$(X, Y, A, P(\theta), Q(\theta), Q_0(\theta), p, r(\theta))$$

with state transition law $P(dx' \mid x,a,\theta)$, observation kernels $Q(dy \mid a,x,\theta)$ and $Q_0(dy \mid x,\theta)$, and reward function $r(x,a,\theta)$ depending measurably on a parameter θ with values in a Borel space Θ. The PO-CM(θ) is supposed to satisfy jointly in θ the requirements in Definition 2.1. Thus $P(dx' \mid x,a,\theta)$ is now a stochastic kernel on X given $XA\Theta$, and similarly for the observation kernels $Q(\theta)$ and $Q_0(\theta)$, whereas $r(x,a,\theta)$ is a function in $B(XA\Theta)$. Our main objective is to give conditions under which the adaptive control results in Section 2.5 hold for partially observable systems.

To begin with, as in Section 4.3, the PO-CM(θ) can be transformed, for each value of θ, into a completely observable control model

5.1 $$\text{CO-CM}(\theta) = (\mathbf{Z}, A, q(\theta), q_0(\theta), r'(\theta)),$$

where the transition law is a stochastic kernel on X given $\mathbf{Z}A\Theta$ defined by

5.2 $$q(D \mid z,a,\theta) := \int_Y 1_D[H(z,a,y,\theta)]\, R'(dy \mid z,a,\theta), \text{ with}$$

5.3 $$H(z,a,y,\theta) := H'(\cdot \mid z,a,y,\theta)$$

and $R'(dy \mid z,a,\theta)$ is defined via 3.3–3.6 when $P(dx' \mid x,a)$ and $Q(dy \mid a,x)$ are replaced by $P(dx' \mid x,a,\theta)$ and $Q(dy \mid a,x,\theta)$, respectively. Similarly, the initial distribution $q_0(dz \mid p,\theta)$ is defined by 3.10, via 3.8 and 3.9, when $Q_0(dy \mid x)$ is replaced by $Q_0(dy \mid x,\theta)$ and $p \in \mathbf{Z} = \mathbf{P}(X)$ is the given *a priori* distribution of the initial state x_0. Finally the one-step reward r' in 5.1 is defined by

5.4 $$r'(z,a,\theta) := \int_X r(x,a,\theta)z(dx) \quad \text{for all} \quad (z,a,\theta) \in \mathbf{Z}A\Theta.$$

Also, the state process $\{z_t\}$ in CO-CM(θ) can be defined by the θ-analog of equations 3.1:

5.5 $$z_{t+1} = H(z_t, a_t, y_{t+1}, \theta), \text{ where } t = 0,1,\ldots, \text{ and } z_0 = H_0(p, y_0, \theta).$$

Now, to obtain the PO-version of the adaptive control results in Section 2.5 we need two sets of assumptions: one set is to obtain the θ-analog of the dynamic programming Theorem 4.3, and the other one is to insure the "continuity" of CO-CM(θ) in θ (which would be the PO-version of Assumption 5.5 in Chapter 2). The former set is, of course, the θ-analog of Assumptions 4.1:

5.6 Assumptions.

(a) A is compact.

(b) $r(x,a,\theta) \in C(XA\Theta)$; let R be a constant such that $|r(x,a,\theta)| \leq R$ for all (x,a,θ).

(c) $P(dx' \mid x,a,\theta)$, $Q(dy \mid a,x,\theta)$ and $Q_0(dy \mid x,\theta)$ are continuous stochastic kernels.

(d) $H(z,a,y,\theta) \in C(\mathbf{Z}AY\Theta)$.

The "continuity" requirements are the following assumptions on PO-CM(θ).

5.7 Assumptions. For any $\theta \in \Theta$ and any sequence $\{\theta_n\}$ in Θ converging to θ, each of the following sequences converges to zero as $n \to \infty$:

$$\rho'(n,\theta) := \sup_{x,a} |r(x,a,\theta_n) - r(x,a,\theta)|,$$

$$\pi'(n,\theta) := \sup_{x,a} \|P(\cdot \mid x,a,\theta_n) - P(\cdot \mid x,a,\theta)\|,$$

and

$$\pi''(n,\theta) := \sup_{a,x} \|Q(\cdot \mid a,x,\theta_n) - Q(\cdot \mid a,x,\theta)\|.$$

Under Assumptions 5.6, we obtain the θ-analog of Lemma 4.2, in particular, $r'(z,a,\theta)$ is a function in $C(\mathbf{Z}A\Theta)$ and

$$\text{5.8} \qquad |r'(z,a,\theta)| \leq R \quad \text{for all} \quad (z,a,\theta),$$

and we also obtain the θ-version of Theorem 4.3. In particular, the optimal reward function $v^*(z,\theta)$ for the CO-CM(θ) is the unique solution in $C(\mathbf{Z}\Theta)$ satisfying, for each $\theta \in \Theta$, the θ-DPE

$$\text{5.9} \quad v^*(z,\theta) = \max_{a \in A} \left\{ r'(z,a,\theta) + \beta \int v^*(z',\theta) q(dz' \mid z,a,\theta) \right\} =: T_\theta v^*(z,\theta).$$

On the other hand, the continuity Assumptions 5.7 are inherited by the completely observable model CO-CM(θ) in the following sense.

5.10 Proposition. *Suppose that Assumptions* 5.6 *and* 5.7 *hold, and that* $\theta_n \to \theta$. *Then:*

(a) $\rho(n,\theta) \leq \rho'(n,\theta) \to 0$, *where*

$$\rho(n,\theta) := \sup_{z,a} |r'(z,a,\theta_n) - r'(z,a,\theta)|.$$

(b) $\pi(n,\theta) \leq 2[\pi'(n,\theta) + \pi''(n,\theta)] \to 0$, *where*

$$\pi(n,\theta) := \sup_{z,a} \|R'(\cdot\,|\,z,a,\theta_n) - R'(\cdot\,|\,z,a,\theta)\|.$$

(c) $\sup_z |v^*(z,\theta_n) - v^*(z,\theta)| \leq c_1 \cdot \max\{\rho(n,\theta), \pi(n,\theta)\}$, *where*

$$c_1 := (1+\beta c_0)/(1-\beta),$$

and

$$c_0 := R/(1-\beta) \geq |v^*(z,\theta)| \quad \textit{for all} \quad (z,\theta) \in \mathbf{Z}\Theta.$$

Note that part (c) is the "same" as Proposition 5.6 in Chapter 2.

Proof. (a) This part follows from Definition 5.4 of r', since

$$\begin{aligned} |r'(z,a,\theta_n) - r'(z,a,\theta)| &\leq \int |r(x,a,\theta_n) - r(x,a,\theta)| z(dx) \\ &\leq \rho'(n,\theta). \end{aligned}$$

(b) By Definition 3.5 of R', we obtain that, for any Borel subset C of Y,

$$|R'(C\,|\,z,a,\theta_n) - R'(C\,|\,z,a,\theta)| \leq \sup_{x\in X} \Big| \int_X Q(C\,|\,a,x',\theta_n)P(dx'\,|\,x,a,\theta_n) - \int_X Q(C\,|\,a,x',\theta)P(dx'\,|\,x,a,\theta)\Big|.$$

Inside the absolute value on the r.h.s. add and subtract the term

$$\int_X Q(C\,|\,a,x',\theta_n)\,P(dx'\,|\,x,a,\theta),$$

and then a straightforward calculation yields that

$$|R'(C\,|\,z,a,\theta_n) - R'(C\,|\,z,a,\theta)| \leq \pi'(n,\theta) + \pi''(n,\theta)$$

for all $C \in \mathcal{B}(Y)$ and (z,a) in $\mathbf{Z}A$. The latter inequality, together with B.2 in Appendix B, implies (b).

Part (c) follows from (a) and (b), exactly as in the proof of Proposition 5.6 [or that of Theorem 4.8(a)] in Chapter 2. □

PEC and NVI I-Policies

From Proposition 5.10 above we can derive (adaptive) θ-ADO I-policies for CO-CM(θ), for every θ in Θ, exactly as in Section 2.5. Specifically, let

$f^* : \mathbf{Z}\Theta \to A$ be a measurable function such that $f^*(z, \theta) \in A$ maximizes the r.h.s. of 5.9 for all (z, θ) in $\mathbf{Z}\Theta$. Then given any sequence $\{\theta_t\}$ in Θ, we consider the I-policy $\delta_\theta^* = \{f_t^*\}$ defined by

$$f_t^*(z) := f^*(z, \theta_t), \quad \text{where} \quad z \in \mathbf{Z}.$$

This is the PEC (or NVI-1) I-policy and, exactly as in Corollary 5.10 of Chapter 2, we obtain:

5.11 If θ_t converges to θ, then δ_θ^* is θ-ADO for CO-CM(θ).

Similarly, we define functions

5.12 $v_t(z, \theta_t) := T_t v_{t-1}(z, \theta_{t-1})$ for all $z \in \mathbf{Z}$ and $t \geq 0$,

where $T_t := T_{\theta_t}$ and $v_{-1}(z) := 0$ for all $z \in \mathbf{Z}$, and let $\overline{\delta}_\theta = \{\overline{f}_t\}$ be the I-policy defined by $\overline{f}_t(z) := f_t(z, \theta_t)$ for all $t \geq 0$, where $f_t(z, \theta_t)$ maximizes the r.h.s. of 5.12 for all $z \in \mathbf{Z}$ and $t \geq 0$. And again as in Corollary 5.10 of Chapter 2, we obtain:

5.13 If $\theta_t \to \theta$, then $\sup_z |v_t(z, \theta_t) - v^*(z, \theta)| \to 0$, and $\overline{\delta}_\theta$ is θ-ADO for CO-CM(θ).

All the other results in Section 2.5 can be translated into results for I-policies for CO-CM(θ), and these in turn, using Propositions 3.11 and 3.13 (in the present chapter), can be re-stated as results for the original partially observable control model PO-CM(θ).

4.6 Comments and References

We have shown in this chapter how one can obtain adaptive (I-)policies for PO-CM's with unknown parameters. This is done in two steps. First, the PO-CM is transformed into a new completely observed (CO) control model (Section 4.3) and conditions are imposed so that the CO-CM satisfies the usual compactness and continuity conditions (Section 4.4). Once we have this, the second step is simply to apply to the CO-CM the results for adaptive control in Chapter 2. This general procedure can be used for PO *semi*-Markov control models and/or problems with average reward criterion [Wakuta (1981, 1982), Acosta Abreu (1987b)]. It can also be used for PO systems of the form 1.1 (see also equations 2.2 and 2.3) when the unknown "parameters" are the distributions μ and ν of the disturbance processes $\{\xi_t\}$ and $\{\eta_t\}$, in which case μ and ν can be estimated (e.g.) using empirical processes (as in Section 2.6); see Hernández-Lerma and Marcus (1989).

The material in Sections 4.2, 4.3 and 4.4 is quite standard; some references are Bertsekas and Shreve (1978), Dynkin and Yushkevich (1979), Striebel (1975), Rhenius (1974), Yushkevich (1976), etc. The main idea (transforming the PO-CM into a CO-CM), however, apparently goes back to papers by Shirjaev or Dynkin in the early 1960's [see references in Dynkin and Yushkevich (1979) or Hinderer (1970)]. Some interesting applications of PO-CM's with finite state space are discussed by Monahan (1982).

Adaptive I-policies in Section 4.5 are reported in Hernández-Lerma and Marcus (1987) for PO-CM's with *denumerable* state space and observation set. To apply these adaptive policies one needs, of course, strongly consistent estimators of the unknown parameters; this problem is discussed in the following chapter. Some works on parameter estimation in non-controlled PO systems are Baum and co-workers (1970) and Loges (1986).

5

Parameter Estimation in MCM's

5.1 Introduction

Let Θ be a Borel space, and for each $\theta \in \Theta$, let $\text{MCM}(\theta) = (X, A, q(\theta), r(\theta))$ be a Markov control model (MCM) as in Section 1.2. Thus X and A are Borel spaces and we assume that the transition law $q(\cdot \mid k, \theta)$ is a stochastic kernel on X given $\mathbf{K}\Theta$, and the one-step reward function $r(k, \theta)$ is a real-valued measurable function on $\mathbf{K}\Theta$. Recall (Definition 2.1 in Chapter 1) that $\mathbf{K}$ is the set of admissible state-action pairs $k = (x, a)$, where $x \in X$ and $a \in A(x)$. We will assume that $A(x)$ is a compact subset of A for every state $x \in X$.

In this chapter, we suppose that the "true" parameter value, say, θ^*, is *unknown;* it is only assumed that θ^* is a (fixed) point in the parameter set Θ. We present below a statistical method to obtain a sequence of "strongly consistent" estimators of θ^* in the sense of the following definition.

1.1 Definition. For each $t = 0, 1, \ldots$, let $\hat{\theta}_t$ be a measurable function from H_t to Θ. (Recall that H_t is the space of histories, h_t, up to time t; see Section 1.2.) It is said that $\{\hat{\theta}_t\}$ is a sequence of *strongly consistent* (SC) *estimators* of a parameter value $\theta \in \Theta$ if, as $t \to \infty$, $\hat{\theta}_t$ converges to θ $P_x^{\delta,\theta}$-a.s. for any policy $\delta \in \Delta$ and any initial state $x \in X$.

Summary

We begin in Section 5.2 by introducing the concept of a contrast function; examples are presented to illustrate how the minimum contrast method, under suitable "identifiability" conditions, includes some commonly used statistical parameter-estimation methods. In Section 5.3, minimum contrast estimators (MCE's) are defined and conditions sufficient for their strong consistency are presented. We close the chapter in Section 5.4 with some brief comments on related results.

It is important to note that for the purposes of the present chapter, the actual performance criterion (for the control problem) is irrelevant.

5.2 Contrast Functions

2.1 Definition. Let $h(\theta, k, x)$ be a real-valued measurable function on $\Theta\mathbf{K}X$, and let

$$H(\theta, \theta', k) := \int_X [h(\theta', k, y) - h(\theta, k, y)]\, q(dy \mid k, \theta).$$

The function h is said to be a *contrast function* if, for every θ and θ' in Θ and $k = (x, a) \in \mathbf{K}$,

(a) $\int_X |h(\theta', k, y)|\, q(dy \mid k, \theta) < \infty$, and

(b) $H(\theta, \theta', k) \geq 0$ and equality holds only if $\theta = \theta'$.

Condition 2.1(b) is called the *identifiability* property of h.

2.2 Example (Conditional Least Squares). Let $v(k, x)$ be a real-valued measurable function on $\mathbf{K}X$ such that

$$\int v^2(k, y)\, q(dy \mid k, \theta) < \infty \quad \text{for all} \quad k \in \mathbf{K} \quad \text{and} \quad \theta \in \Theta,$$

and define the mean-value function

$$m(\theta, k) := \int v(k, y)\, q(dy \mid k, \theta).$$

Suppose that $m(\theta, k) \neq m(\theta', k)$ for all $k \in \mathbf{K}$ if $\theta \neq \theta'$. Then the function

$$h(\theta, k, y) := [v(k, y) - m(\theta, k)]^2$$

is a contrast function.

Indeed, 2.1(a) holds by the assumption on v, whereas to prove 2.1(b) we note that the function $H(\theta, \theta', k)$ can be written as

$$\begin{aligned} H(\theta, \theta', k) = & \int [m(\theta, k) - m(\theta', k)][2v(k, y) - m(\theta, k) - m(\theta', k)]\, q(dy \mid k, \theta) \\ = & \ [m(\theta, k) - m(\theta', k)][2m(\theta, k) - m(\theta, k) - m(\theta', k)] \\ = & \ [m(\theta, k) - m(\theta', k)]^2 \geq 0, \end{aligned}$$

and 2.1(b) follows.

In particular, if we take the function v above as $v(k, y) := y$, then we can write h as

$$h(\theta, x_t, a_t, x_{t+1}) = [x_{t+1} - E(x_{t+1} \mid x_t, a_t, \theta)]^2.$$

The minimum contrast estimators (Definition 3.1 below) defined in terms of such a function h are called "conditional least squares" or "prediction error" estimators; see, e.g., Klimko and Nelson (1978), or Ljung (1981). In Ljung's terminology, the term in parenthesis,

$$x_{t+1} - E(x_{t+1} \mid x_t, a_t, \theta)$$

is called the "prediction error" or "residual". These estimators are also used by Loges (1986) in partially observable linear systems.

2.3 Example (Maximum Likelihood). Suppose $q(\cdot \mid k, \theta)$ has a strictly positive transition density $p(\theta, k, \cdot)$ with respect to some sigma-finite measure μ on X; that is, $p(\theta, k, y)$ is a positive measurable function on $\Theta \mathbf{K} X$ such that

$$q(B \mid k, \theta) = \int_B p(\theta, k, y)\mu(dy) \quad \text{for every} \quad B \in \mathcal{B}(X).$$

Suppose that, for all θ and θ' in Θ and k in $\mathbf{K}$,

$$\int |\mathrm{Log}\, p(\theta', k, y)|\, q(dy \mid k, \theta) < \infty,$$

and, moreover,

$$\textbf{2.4} \qquad q(\cdot \mid k, \theta) \neq q(\cdot \mid k, \theta') \quad \text{for all } k \in \mathbf{K} \text{ if } \theta \neq \theta'.$$

Then the function

$$h(\theta, k, y) := -\mathrm{Log}\, p(\theta, k, y)$$

is a contrast function.

Indeed, the integrability condition 2.1(a) holds (by assumption), whereas 2.1(b) follows from Jensen's inequality (Lemma 2.6 below). That is, the function $H(\theta, \theta', k)$ in Definition 2.1 becomes

$$\textbf{2.5} \qquad \begin{aligned} H(\theta, \theta', k) &= \int \mathrm{Log}[p(\theta, k, y)/p(\theta', k, y)] \cdot p(\theta, k, y)\mu(dy) \\ &= \int L(y) \cdot (\mathrm{Log}\, L(y)) \cdot p(\theta', k, y)\, \mu(dy), \end{aligned}$$

where $L(y) := p(\theta, k, y)/p(\theta', k, y)$. Therefore, since the function $f(t) = t\, \mathrm{Log}\, t$ is strictly convex in $t > 0$, we obtain

$$\begin{aligned} H(\theta, \theta', k) &\geq \left[\int L(y)\, p(\theta', k, y)\mu(dy)\right] \cdot \mathrm{Log}\left[\int L(y)\, p(\theta', k, y)\mu(dy)\right] \\ &= \left[\int p(\theta, k, y)\mu(dy)\right] \cdot \mathrm{Log}\left[\int p(\theta, k, y)\mu(dy)\right] = 0. \end{aligned}$$

Actually, by Assumption 2.4 and the strict convexity of $f(t)$, the latter inequality is strict if $\theta \neq \theta'$.

The function $H(\theta, \theta', k)$ in 2.5 is called Kullback's information of $q(\cdot \mid k, \theta)$ relative to $q(\cdot \mid k, \theta')$ [see, e.g., Maigret (1979), or Hijab (1987), p. 78], and the minimum contrast estimators (Definition 3.1) defined in terms of $-\text{Log}\, p(\theta, k, y)$ are called "maximum likelihood" estimators.

Examples 2.2 and 2.3 show that some of the standard parameter-estimation methods can be studied in terms of contrast functions, provided that appropriate identifiability conditions (such as 2.4 above) hold. As another example, see Ivanov and Kozlov (1981) for a discussion of Wolfowitz' (1957) minimum-distance estimation method in terms of contrast functions.

2.6 Lemma. [Kolonko (1982)]. *Let x be an integrable random variable taking values in an open interval I, and let f be a twice-differentiable function on I, with $f'' > 0$ on I. Assume further that $E|f(x)| < \infty$. Then:*

(a) $Ef(x) \geq f(Ex)$, *and*

(b) $Ef(x) = f(Ex)$ *if and only if x is constant a.s.*

Proof. Let $c := E(x)$ be the expected value of x; c is in I. Using Taylor's expansion we obtain

$$Ef(x) = f(c) + \frac{1}{2}E\{(x-c)^2 \cdot f''[\alpha(x) \cdot (x-c) + c\},$$

for some function α from I to $(0, 1)$. Then (a) follows since $f'' > 0$. On the other hand, $Ef(x) = f(c)$ if and only if

$$E\{(x-c)^2 \cdot f''[\alpha(x) \cdot (x-c) + c]\} = 0,$$

or equivalently,

$$(x-c)^2 \cdot f''[\alpha(x) \cdot (x-c) + c] = 0 \quad \text{a.s.}$$

Since $f'' > 0$, the latter is equivalent to $x = c$ a.s. □

Lemma 2.6 is a strong version of Jensen's inequality.

5.3 Minimum Contrast Estimators

Throughout the following we consider an adaptive $\text{MCM}(\theta) = (X, A, q(\theta), r(\theta))$ for which the "true" parameter value is an arbitrary (fixed) point θ^* in Θ. Sometimes we write $q(\cdot \mid k, \theta^*)$ as $q^*(\cdot \mid k)$; similarly, for any policy δ and any initial state $x_0 = x$, we write

$$P_x^{\delta,\theta^*} \text{ and } E_x^{\delta,\theta^*} \text{ as } P_x^{*\delta} \text{ and } E_x^{*\delta},$$

respectively.

Our purpose in this section is to use the method of minimum contrast estimation (explained below) to show the existence of a sequence of strongly consistent (SC) estimators of θ^*.

Let $h(\theta, k, y)$ be a given (fixed) contrast function and, for each $t \geq 1$, let L_t be the function on ΘH_t defined by

$$L_t(\theta, h_t) := \sum_{j=0}^{t-1} h(\theta, x_j, a_j, x_{j+1}),$$

where $h_t = (x_0, a_0, \ldots, x_{t-1}, a_{t-1}, x_t)$. Sometimes we write $L_t(\theta, h_t)$ simply as $L_t(\theta)$.

3.1 Definition. A measurable function $\hat{\theta}_t : H_t \to \Theta$ such that $\hat{\theta}_t = \hat{\theta}_t(h_t)$ minimizes $L_t(\theta, h_t)$, that is,

$$L_t(\hat{\theta}_t, h_t) = \min_{\theta \in \Theta} L_t(\theta, h_t) \quad \text{for all} \quad h_t \in H_t,$$

is said to be a *minimum contrast estimator* (MCE) of θ^*.

To insure the existence of such estimators we suppose the following.

3.2 Assumption. Θ is a compact metric space, and $h(\theta, k, y)$ is a continuous function of $\theta \in \Theta$ for all (k, y) in $\mathbf{K}X$.

This assumption, together with the measurability of $L_t(\theta, h_t)$, implies the existence of a measurable minimizer $\hat{\theta}_t$ of L_t. (To see this, simply translate the "maximization" in Proposition D.3, Appendix D, into a suitable minimization problem.)

Let us now consider the following statements 3.3, 3.4, and 3.5.

3.3 X and A are compact. Moreover,

(a) the contrast function $h(\theta, k, y)$ is continuous on $\Theta \mathbf{K} X$, and

(b) $q^*(\cdot \mid k) = q(\cdot \mid k, \theta^*)$ is (weakly-) continuous in $k \in \mathbf{K}$; that is, for any continuous bounded function v on X, the integral

$$\int v(y) q^*(dy \mid k) \quad \text{is continuous in} \quad k.$$

3.4 X and A are Borel spaces (as usual). Moreover,

(a) The family of functions $\{h(\cdot, k, y) \mid (k, y) \in \mathbf{K}\}$ is equicontinuous at each point $\theta \in \Theta$.

(b) For every open neighborhood U of θ^*,

$$\inf\{H^*(\theta, k) \mid \theta \notin U \text{ and } k \in \mathbf{K}\} > 0,$$

where $H^*(\theta, k) := H(\theta^*, \theta, k)$ is the function in Definition 2.1.

3.5 There exists a sequence $\{\hat{\theta}_t\}$ of MCE's of θ^*, and $\{\hat{\theta}_t\}$ is a sequence of SC estimators of θ^* (Definition 1.1).

We now state the main result.

3.6 Theorem. *Suppose that Assumption* 3.2 *holds and that the set* **K** *is closed. Suppose also that the contrast function h satisfies that*

$$\int h^2(\theta,k,y)\, q^*(dy\,|\,k) \le C < \infty \quad \textit{for all} \quad (\theta,k) \in \Theta\mathbf{K}.$$

Then 3.3 *implies* 3.4, *and* 3.4 *implies* 3.5.

A sufficient condition for **K** to be closed is given in Proposition D.3(d), Appendix D.

Proof. 3.3 *implies* 3.4: Suppose that 3.3 holds. Then part 3.4(a) is obvious. To prove 3.4(b), let us first note that the function $H^*(\theta,k)$ is continuous on $\Theta\mathbf{K}$ (by Proposition C.2(b), Appendix C). Thus, for any open neighborhood U of θ^*, $H^*(\theta,k)$ attains its minimum value on the compact set $U^c\mathbf{K}$, where U^c denotes the complement of U in Θ, and such a value is strictly positive, by definition of contrast function. This completes the proof that 3.3 implies 3.4.

3.4 *implies* 3.5: Under the present assumptions, there exists a sequence $\{\hat{\theta}_t\}$ of MCE's of θ^*. We shall now prove that $\{\hat{\theta}_t\}$ is strongly consistent, i.e., given any open neighborhood U of θ^*,

$$\hat{\theta}_t \in U \;\; P_x^{*\delta}\text{-a.s. for all } t \text{ sufficiently large}$$

for any policy δ and any initial state x.

Let us fix U, δ and x, and write $P_x^{*\delta} = P^*$ and $E_x^{*\delta} = E^*$.

Let D be a countable dense subset of Θ. We shall use the following definitions:

$$h^*(\theta,k,y) := h(\theta,k,y) - h(\theta^*,k,y),$$

$$V_n(\theta) := \{\theta' \in \Theta \,|\, d(\theta,\theta') \le 1/n\},$$

where d denotes the metric on Θ, and for any closed ball B in Θ we write

$$h(B,k,y) := \inf_{\theta\in B} h(\theta,k,y)$$

and

$$h^*(B,k,y) := h(B,k,y) - h(\theta^*,k,y).$$

Notice that $H^*(\theta,k) = \int h^*(\theta,k,y)\, q^*(dy\,|\,k)$, and, on the other hand,

$$h(B,k,y) = h(B\cap D,k,y),$$

so that $h(B,k,y)$ is measurable in (k,y) for every closed ball B.

Now, by 3.4(a), given any positive number ϵ, there is an integer n such that, if $d(\theta, \theta') \leq 1/n$, then

$$|h(\theta', k, y) - h(\theta, k, y)| \leq \epsilon \text{ for all } (k, y) \in \mathbf{K}X,$$

so that

3.7 $|h^*(V_n(\theta), k, y) - H^*(\theta, k, y)| \leq \epsilon$ for all $\theta \in \Theta$ and $(k, y) \in \mathbf{K}X$.

Observe that, as $n \to \infty$, $V_n(\theta)$ decreases to $\{\theta\}$, and $h^*(V_n(\theta), k, y)$ increases to $h^*(\theta, k, y)$. Therefore, if in 3.4(b) we let $\eta > 0$ be such that

$$\inf_{\theta \notin U, k \in \mathbf{K}} H^*(\theta, k) = 2\eta$$

and take $\epsilon = 2\eta$ in 3.7, then for all n sufficiently large

3.8 $$\inf_{\theta \notin U, k \in \mathbf{K}} \int h^*(V_n(\theta), k, y)\, q^*(dy \mid k) \geq \eta.$$

Now, given $\theta_i \in D \cap U^c$ and $n \geq 1$, define the random variables

$$u_t := h^*(V_n(\theta_i), x_t, a_t, x_{t+1}) - \int h^*(V_n(\theta_i), x_t, a_t, y)\, q^*(dy \mid x_t, a_t)$$

and

$$M_t := \sum_{s=0}^{t-1} u_s.$$

u_t is $\mathcal{F}_{t+1}$-measurable, where $\mathcal{F}_t$ is the sigma-algebra generated by the history h_t, and moreover, M_t is a martingale, since

$$E^*(M_{t+1} - M_t \mid \mathcal{F}_t) = E^*(u_t \mid \mathcal{F}_t) = 0.$$

On the other hand (by the assumptions),

$$\sum_t t^{-2} E^*(u_t^2 \mid \mathcal{F}_t) < \infty,$$

whence, by the Law of Large Numbers for martingales (Lemma 3.11 below),

3.9 $$n^{-1} M_n \to 0 \quad P^*\text{-a.s. as } n \to \infty.$$

Therefore, since

$$\inf_i t^{-1} \sum_{s=0}^{t-1} \int h^*(V_n(\theta_i), x_s, a_s, y)\, q^*(dy \mid x_s, a_s) \geq \eta,$$

the limit in 3.9 implies

$$\liminf_t \inf_i t^{-1} \sum_{s=0}^{t-1} h^*(V_n(\theta_i), x_s, a_s, x_{s+1}) \geq \eta \quad P^*\text{-a.s.},$$

which is equivalent to

$$\liminf_{t} \inf_{\theta \notin U} t^{-1} \sum_{s=0}^{t-1} h^*(\theta, x_s, a_s, x_{s+1}) \geq \eta \quad P^*\text{-a.s.}$$

This implies that for all t sufficiently large

$$\inf_{\theta \notin U} t^{-1} \sum_{s=0}^{t-1} h^*(\theta, x_s, a_s, x_{s+1})$$

$$= \inf_{\theta \notin U} t^{-1} \sum_{s=0}^{t-1} [h(\theta, x_s, a_s, x_{s+1}) - h(\theta^*, x_s, a_s, x_{s+1})]$$

$$\textbf{3.10} \qquad = \inf_{\theta \notin U} t^{-1} [L_t(\theta) - L_t(\theta^*)] > \eta/2 \quad P^*\text{-a.s.}$$

On the other hand, by Definition 3.1 of MCE, $L_t(\hat{\theta}_t) \leq L_t(\theta^*)$ for all t, that is,

$$t^{-1} \sum_{s=0}^{t-1} [h(\hat{\theta}_t, x_s, a_s, x_{s+1}) - h(\theta^*, x_s, a_s, x_{s+1})] \leq 0.$$

Therefore, necessarily $\hat{\theta}_t \in U$ P^*-a.s. for all t sufficiently large, since otherwise we obtain a contradiction to 3.10. This completes the proof of Theorem 3.6. □

3.11 Lemma. *Let $\{M_n := \sum_{t=1}^{n} y_t, \mathcal{F}_n, n \geq 1\}$ be a martingale such that*

$$\sum_{t=1}^{\infty} t^{-2} E(y_t^2 \mid \mathcal{F}_{t-1}) < \infty,$$

where $\mathcal{F}_0$ is the trivial sigma-algebra. Then

$$\lim_{n \to \infty} n^{-1} M_n = 0 \quad a.s.$$

Proof. Hall and Heyde (1980), p. 35; Loève (1978), p. 53.

5.4 Comments and References

The minimum contrast method for parameter estimation goes back to Huber (1967) and Pfanzagl (1969); see also Gänssler (1972). It was first used in adaptive control problems by Mandl (1974), who considered finite-state controlled Markov chains; it was extended to denumerable-state Markov renewal models by Kolonko (1982), and to MCM's with general state space by Georgin (1978b). Our presentation here is borrowed mainly from the latter reference.

Additional properties of the MCE's are studied by Mandl (1974) and Maigret (1979). For instance, using a large deviations result, Maigret shows *exponential* consistency of the MCE's in Theorem 3.6 above, if condition 3.3 holds.

Other approaches to parameter estimation in stochastic control systems are given, e.g., by Borkar and Varaiya (1982), Doshi and Shreve (1980), El-Fattah (1981), Kumar (1985), Kurano (1972, 1983, 1985) and Ljung (1981). On the other hand, in some adaptive control problems it is sometimes possible to use elementary estimation procedures, e.g., moment estimators. This is the case, in particular, in control of queueing systems, in which typically the unknown parameters are mean values (say, mean arrival rates and/or mean service times) and strong consistency can be proved using appropriate variants of the Law of Large Numbers: see (e.g.) Cavazos-Cadena and Hernández-Lerma (1987), Hernández-Lerma and Marcus (1983, 1984), Kolonko (1982), and references therein.

For remarks and references on parameter estimation in partially observable MCM's see Hernández-Lerma and Marcus (1987).

6

Discretization Procedures

6.1 Introduction

In previous chapters we have studied adaptive and non-adaptive Markov control problems within the framework of stochastic Dynamic Programming (DP), and we have also shown in Chapters 2 and 3 several ways to approximate the optimal reward function, denoted by v^*. There are many cases, however, in which it is difficult to get either v^* or the approximating functions in closed form, and one is thus led to consider other types of approximations, suitable for numerical implementation.

We consider in this chapter the standard state-space discretization, and show how it can be extended to yield recursive approximations to adaptive and non-adaptive Markov control problems with a discounted reward criterion.

Summary

We begin in Section 6.2 by introducing the control model (CM) we will be dealing with, together with some general assumptions. In Section 6.3 we consider non-adaptive CM's. We start with a non-recursive procedure and then show how it can be made recursive; furthermore, in both cases, recursive and non-recursive, we obtain asymptotically discount optimal (ADO) control policies. The results in Section 6.3 can be seen as "discretized" versions of the Nonstationary Value Iteration (NVI) schemes NVI-1 and NVI-2 in Section 2.4. Next, in Section 6.4, the discretizations in Section 6.3 are extended to adaptive CM's and, in particular, we obtain "discretized" forms of the Principle of Estimation and Control (PEC) and the NVI adaptive policies in Section 2.5. The proofs of the theorems in Sections 6.3 and 6.4 have many common arguments, and therefore, all the proofs are collected in Section 6.5. We close in Section 6.6 with some comments on relevant references.

6.2 Preliminaries

We consider the usual Markov control model CM $= (X, A, q, r)$ defined in Section 1.2 (Definition 2.1), but in addition we will assume that X *and* A *are compact metric spaces* with metrics d_1 and d_2, respectively. The optimal control problem is to find a policy that maximizes the expected total discounted reward (cf. Section 2.1)

$$V(\delta, x) := E_x^\delta \sum_{t=0}^{\infty} \beta^t r(x_t, a_t).$$

As usual, the optimal reward function is denoted by v^*:

$$v^*(x) := \sup_\delta V(\delta, x), \quad x \in X.$$

For the convergence of the discretization procedures to be introduced below, we need v^* to be a Lipschitz function, and therefore, we make here the same assumptions leading to Lemma 6.8 in Section 2.6.

2.1 Assumptions. Let $d := \max\{d_1, d_2\}$ be a metric on $\mathbf{K} := \{(x,a) \mid x \in X, a \in A(x)\}$, and suppose there are constants R, L_0, L_1, and L_2, such that

(a) $|r(k)| \leq R$ and $|r(k) - r(k')| \leq L_0 d(k, k')$ for all k and k' in $\mathbf{K}$;

(b) For each $x \in X$, the set $A(x)$ of admissible controls in state x is a closed subset of A, and

$$H(A(x), A(x')) \leq L_1 d_1(x, x') \quad \text{for all } x \text{ and } x' \text{ in } X,$$

where H is the Hausdorff metric (Appendix D);

(c) $\|q(\cdot \mid k) - q(\cdot \mid k')\| \leq L_2 d(k, k')$ for all k and k' in $\mathbf{K}$.

For a discussion of these assumptions see the remarks at the end of Section 2.6. On the other hand, in Section 2.6, it can also be seen that Assumptions 2.1 imply that the optimal reward function v^* is the unique solution in $C(X)$ of the Dynamic Programming Equation (DPE)

2.2 $$v^*(x) = \max_{a \in A(x)} \left\{ r(x,a) + \beta \int v^*(y)\, q(dy \mid x, a) \right\},$$

and, moreover (see Theorem 6.7 or Lemma 6.8 in Section 2.6),

2.3 $$|v^*(x) - v^*(x')| \leq L^* \cdot d_1(x, x') \quad \text{for all } x \text{ and } x' \text{ in } X,$$

where

2.4 $$L^* := (L_0 + \beta L_2 c_0) \cdot \max\{1, L_1\}, \text{ with } c_0 := R/(1-\beta).$$

Assumptions 2.1 are assumed to hold throughout.

6.3 The Non-Adaptive Case

We will now introduce two discretization procedures for the Markov control model CM $= (X, A, q, r)$ in Section 6.2. The idea is very simple: we introduce Markov control models $\mathrm{CM}_n = (X_n, A_n, q_n, r_n)$ with a *finite* state space X_n and such that, as $n \to \infty$, CM_n approximates, in some sense, the original control model CM.

A Non-Recursive Procedure

We begin by partitioning the state space X as follows.

Partition of X. For each $n = 0, 1, \ldots$, let $P_n := \{X_1^n, \ldots, X_{m_n}^n\}$ be a partition of X; that is, the sets X_i^n, for $i = 1, \ldots, m_n$, are disjoint (nonempty) measurable subsets of X whose union is X. For each $i = 1, \ldots, m_n$, select an arbitrary point x_i^n in X_i^n, and call X_n the set consisting of all those points. We define the *diameter* of P_n as

3.1 $$D(n) := \sup_{1 \le i \le m_n} \sup_{x \in X_i^n} d_1(x, x_i^n).$$

We will assume that the partitions are chosen so that, for every $n \ge 0$,

- P_{n+1} is a "refinement" of P_n; that is, every set in P_n is a union of sets in P_{n+1};
- X_{n+1} contains X_n; and
- the diameters $D(n)$ form a nonincreasing sequence that converges to zero.

Construction of CM_n. Given the set $X_n = \{x_1^n, \ldots, x_{m_n}^n\}$, sometimes called the "grid", define:

3.2 (a) the control set $A_n := \cup_{i=1}^{m_n} A(x_i)$.

(b) the transition law $q_n(x_j^n \mid x, a) := q(X_j^n \mid x, a)$ for $j = 1, \ldots, m_n$, and $(x, a) \in K_n$, where $K_n := \{(x, a) \mid x \in X_n \text{ and } a \in A(x)\}$. (Note that q_n is a stochastic kernel on X_n given K_n.)

(c) the one-step reward function $r_n(x, a) := r(x, a)$ for $(x, a) \in K_n$.

These sets and functions define a Markov control model $\mathrm{CM}_n := (X_n, A_n, q_n, r_n)$ for a controlled Markov chain with *finite* state space X_n.

Associated with CM_n we consider a discounted reward problem with the same discount factor β as for the original problem for CM. Thus from the results in Section 2.2 (Theorem 2.2), the optimal reward function for CM_n, denoted by $v_n^*(x)$, satisfies the DPE (cf. 2.2)

3.3 $$v_n^*(x) = \max_{a \in A(x)} \left\{ r_n(x, a) + \beta \sum_{j=1}^{m_n} v_n^*(x_j^n)\, q_n(x_j^n \mid x, a) \right\} \quad \text{for } x \in X_n,$$

and, furthermore, a stationary policy f_n^* for CM$_n$ (i.e., a function f_n^* from X_n to A_n such $f_n^*(x) \in A(x)$ for all $x \in X_n$) is optimal if and only if $f_n^*(x)$ maximizes the r.h.s. of 3.3. To compare v_n^* to v^*, we extend v_n^* (and f_n^*) from X_n to X as follows.

Extension of v_n^* and f_n^*. Define v_n^* on X as

3.4 $$v_n^*(x) := v_n^*(x_i^n) \quad \text{for } x \in X_i^n \text{ and } i = 1, \ldots, m_n.$$

In general, we can *not* extend f_n^* to X in the same form, because the function $f : X \to A$ defined by $f(x) := f_n^*(x_i^n)$ for $x \in X_i^n$ might *not* be an admissible stationary policy for CM; that is, the requirement $f(x) \in A(x)$ may fail. We thus proceed differently.

Let $g : \mathbf{K} \to \mathbf{R}$ be the function defined for $(x, a) \in \mathbf{K}$ by

$$g(x, a) := d_2(a, f_n^*(x_i^n)) \quad \text{if } x \in X_i^n \text{ and } i = 1, \ldots, m_n.$$

This function is measurable and it is continuous in $a \in A(x)$ for each $x \in X$. Therefore, by the Measurable Selection Theorem in Proposition D.3 (Appendix D), there exists a stationary policy (or measurable selector) $f_n^* \in \mathbf{F}$, that is, a measurable function $f_n^* : X \to A$ such that $f_n^*(x) \in A(x)$ for all $x \in X$, and

$$\begin{aligned} g(x, f_n^*(x)) &= \min_{a \in A(x)} g(x, a) \quad \text{for every } x \in X \\ &= \min_{a \in A(x)} d_2(a, f_n^*(x_i^n)) \quad \text{if } x \in X_i^n,\ i = 1, \ldots, m_n. \end{aligned}$$

In other words, if $x \in X_i^n$, then $f_n^*(x)$ is a control in $A(x)$ which is "closest" to $f_n^*(x_i^n)$, that is,

3.5 $$d_2[f_n^*(x), f_n^*(x_i^n)] = \text{distance}[f_n^*(x_i^n), A(x)].$$

If there is more than one point $f_n^*(x) \in A(x)$ satisfying 3.5, we can choose any one of them as long as the measurability condition is preserved.

Throughout the following, v_n^* and f_n^* denote the functions on X defined by 3.4 and 3.5, respectively. Notice that, using 3.2(b), we can write the DPE 3.3 as

3.6 $$v_n^*(x) = \max_{a \in A(x)} \left\{ r(x, a) + \beta \int v_n^*(y)\, q(dy \mid x, a) \right\} \quad \text{for } x \in X_n,$$

which will facilitate the comparison with the DPE 2.2. On the other hand, 3.5 together with Assumption 2.1(b) implies that, if $x \in X_i^n$, then

$$d_2[f_n^*(x), f_n^*(x_i^n)] \leq H(A(x), A(x_i^n)) \leq L_1 d_1(x, x_i^n),$$

and therefore, by definition of the diameter $D(n)$,

3.7 $$d_2[f_n^*(x), f_n^*(x_i^n)] \leq L_1 \cdot D(n)$$

for all $x \in X_i^n$, $1 \leq i \leq m_n$, and $n \geq 0$. This in turn, together with the definition of the metric $d := \max\{d_1, d_2\}$ on $\mathbf{K}$, implies

3.8 $$d[(x, f_n^*(x)), (x_i^n, f_n^*(x_i^n))] \leq \max\{1, L_1\} \cdot D(n),$$

for all $x \in X_i^n$, $1 \leq i \leq m_n$, and $n \geq 0$.

Now we can interpret the index $n = 0, 1, \ldots$ in two different ways: on the one hand, it simply denotes the nth discretization (or approximation) of the Markov control model CM $= (X, A, q, r)$; and on the other hand, we can view n as a *time* index, in which case v_n^* is the approximation to v^* obtained *at time n*, and the sequence $\{f_0^*, f_1^*, \ldots\}$ defines a *Markov policy* δ^* for CM that applies the control action $a_n := f_n^*(x_n)$ at time $n = 0, 1, \ldots$. In either case, we have the following.

3.9 Theorem. *For every* $n = 0, 1, \ldots$,

(a) $\|v_n^* - v^*\| \leq c^* D(n)$, *where* $c^* := L^*/(1-\beta)$ *and* L^* *is the Lipschitz constant in* 2.4.

(b) $\sup_x |\phi(x, f_n^*(x))| \leq 2c^* D(n)$, *where* $\phi(x, a)$ *is the discrepancy function defined in Section* 2.3, *i.e.*,

$$\phi(x, a) := r(x, a) + \beta \int v^*(y)\, q(dy \mid x, a) - v^*(x).$$

(c) $\sup_x |V_n(\delta^*, x) - E_x^{\delta^*} v^*(x_n)| \leq 2(1-\beta)^{-1} c^* D(n)$, *where (see Section* 2.1)

$$V_n(\delta, x) := E_x^\delta \sum_{t=n}^{\infty} \beta^{t-n} r(x_t, a_t) \quad \textit{for any policy} \quad \delta, \textit{and } x \in X.$$

In particular, for $n = 0$, *we have* $V_0(\delta, x) = V(\delta, x)$ *and* (c) *becomes*

(d) $\sup_x |V(\delta^*, x) - v^*(x)| \leq 2(1-\beta)^{-1} c^* D(0)$.

(e) *The Markov policy* $\delta^* = \{f_n^*\}$ *is asymptotically discount optimal* (ADO) *for the* CM $= (X, A, q, r)$.

From the results in Section 2.3 (see equation 3.5 and Theorem 3.6), parts (c)–(e) in the above theorem follow from (b). The proofs of (a) and (b) are given in Section 6.5.

The discretization procedure 3.1–3.5 above is similar to the Nonstationary Value Iteration (NVI) scheme NVI-1 in Section 2.4, and as such, it has the inconvenience of not being recursive. We will now turn to a recursive procedure, similar to the scheme NVI-2.

A Recursive Procedure

We consider again the same discretizations P_n, $D(n)$, and CM$_n = (X_n, A_n, q_n, r_n)$ introduced above, with $n = 0, 1, \ldots$, and define recursively a sequence of functions v_n' as follows.

For every $n \geq 0$ and $x \in X_n$, let

3.10 (a) $$v'_n(x) := \max_{a\in A(x)} \left\{ r(x,a) + \beta \sum_{j=1}^{m_n} v'_{n-1}(x_j^n)\, q_n(x_j^n \mid x, a) \right\},$$

where $v'_{-1}(\cdot) \equiv 0$, so that the initial function v'_0 is simply

3.10 (b) $$v'_0(x) := \max_{a\in A(x)} r(x,a), \quad x \in X_0.$$

We then extend v'_n to all of X defining

$$v'_n(x) := v'_n(x_i^n) \text{ for } x \in X_i^n \text{ and } i = 1, \ldots, m_n.$$

Next, we introduce a Markov policy $\delta' := \{f'_0, f'_1, \ldots\}$ for the control model CM $= (X, A, q, r)$ as was done above with δ^*. Namely, for each $n \geq 0$, we define $f'_n : X_n \to A_n$ as a function such that $f'_n(x) \in A(x)$ maximizes the r.h.s. of 3.10(a) for all $x \in X_n$, and then f'_n is extended to a function $f'_n \in \mathbf{F}$ such that,

3.11 $$d_2[f'_n(x), f'_n(x_i^n)] = \min_{a\in A(x)} d_2[a, f'_n(x_i^n)]$$

for all $x \in X_i^n$ and $i = 1, \ldots, m_n$. The inequalities 3.7 and 3.8 also hold, of course, when f_n^* is replaced by f'_n. With this notation, the recursive version of Theorem 3.9 becomes as follows.

3.12 Theorem. *For every* $n = 0, 1, \ldots,$

(a) $\|v'_n - v^*\| \leq c_2 I(n)$, *where* $c_2 = c^* + 2c_0 = (L^* + 2R)/(1-\beta)$,

$$I(n) := \max\{\beta^{[n/2]}, D([n/2])\},$$

and $[c]$ *denotes the largest integer* $\leq c$.

(b) $\sup_x |\phi(x, f'_n(x))| \leq L^* D(n) + \|v'_n - v^*\| + \beta \|v'_{n-1} - v^*\| \leq I'(n)$, *where*

$$I'(n) := L^* D(n) + (1+\beta) c_2 I(n-1) \quad \textit{for all } n \geq 0,$$

with $I(-1) := I(0)$.

(c) $\sup_x |V_n(\delta', x) - E_x^{\pi'} v^*(x_n)| \leq (1-\beta)^{-1} I'(n)$.

(d) $\delta' = \{f'_n\}$ *is* ADO *for* CM $= (X, A, q, r)$.

Once again, as in Theorem 3.9, it can be seen that parts (c) and (d) in Theorem 3.12 follow from the results in Section 2.3. Parts (a) and (b) are proved in Section 6.5.

We have thus obtained *recursive* approximations v'_n to v^*, but the price we pay is that the convergence rates are slower than in Theorem 3.9. On the other hand, as already noted, the discretizations introduced above can be seen as discretized forms of the approximation schemes NVI-1 and NVI-2 in Section 2.4. Of course, we can discretize the scheme NVI-3, but since the approach should be clear by now, we leave the details to the reader.

6.4 Adaptive Control Problems

We now consider adaptive Markov control models CM(θ) $= (X, A, q(\theta), r(\theta))$ with transition law $q(\cdot \mid k, \theta)$ and one-step reward function $r(k, \theta)$ depending on a parameter $\theta \in \Theta$, where Θ is a Borel space. The objective of this section is to extend the discretization procedures in Section 6.3 to the adaptive case. Or, alternatively, the results below can be seen as discretizations of the "Principle of Estimation and Control (PEC)" and the "Nonstationary Value Iteration (NVI)" adaptive policies in Section 2.5.

Preliminaries

We begin by noting (as in Section 2.5) that all functions concerning CM(θ) will be indexed by $\theta \in \Theta$. Thus

$$q(\cdot \mid k),\ r(k),\ V(\delta, x),\ V_n(\delta, x),\ v^*(x),\ \text{etc.},$$

will now be written as

$$q(\cdot \mid k, \theta),\ r(k, \theta),\ V(\delta, x, \theta),\ V_n(\delta, x, \theta),\ v^*(x, \theta),\ \text{etc.}$$

We also need two sets of assumptions. The first one is simply the θ-analogue of Assumptions 2.1, so that for each (fixed) value of θ we have a control model in the context of Sections 6.2 and 6.3.

4.1 Assumptions. (X, d_1) and (A, d_2) are compact metric spaces, and

(a) $r(k, \theta)$ is a measurable function on $\mathbf{K}\Theta$ such that

$$|r(k, \theta)| \leq R, \text{ and } |r(k, \theta) - r(k', \theta)| \leq L_0(\theta) d(k, k'), \text{ with } L_0(\theta) \leq L_0,$$

for all k and k' in $\mathbf{K}$, and $\theta \in \Theta$;

(b) Same as 2.1(b);

(c) $q(\cdot \mid k, \theta)$ is a stochastic kernel on X given $\mathbf{K}\Theta$ such that

$$\|q(\cdot \mid k, \theta) - q(\cdot \mid k', \theta)\| \leq L_2(\theta)\, d(k, k'), \text{ with } L_2(\theta) \leq L_2,$$

for all k and k' in $\mathbf{K}$, and $\theta \in \Theta$.

Under these assumptions, for each value of θ, the optimal reward function $v^*(x, \theta)$ satisfies the θ-DPE in Section 2.5, Theorem 5.2, i.e.,

4.2 $$v^*(x, \theta) = \max_{a \in A(x)} \left\{ r(x, a, \theta) + \beta \int v^*(y, \theta)\, q(dy \mid x, a, \theta) \right\},$$

and it also satisfies the Lipschitz condition 2.3.

The second set of assumptions (again, as in Section 2.5) is about the continuity of the mapping $\theta \to$ CM(θ).

4.3 Assumptions. For any $\theta \in \Theta$ and any sequence $\{\theta_n\}$ in Θ that converges to θ, both sequences $\rho(n,\theta)$ and $\pi(n,\theta)$ converge to zero as $n \to \infty$, where

$$\rho(n,\theta) := \sup_k |r(k,\theta_n) - r(k,\theta)|$$

and

$$\pi(n,\theta) := \sup_k \|q(\cdot \mid k,\theta_n) - q(\cdot \mid k,\theta)\|.$$

We also define $\overline{\rho}(n,\theta) := \rho(n,\theta)$ if the sequence $\rho(n,\theta)$ is nonincreasing, and otherwise,

$$\overline{\rho}(n,\theta) := \sup_{t \geq n} \rho(t,\theta).$$

$\overline{\pi}(n,\theta)$ is defined similarly, and note that $\overline{\rho}(n,\theta) \to 0$ if and only if $\rho(n,\theta) \to 0$, and similarly for $\pi(n,\theta)$ and $\overline{\pi}(n,\theta)$.

Next we extend the discretization 3.1–3.5 to CM(θ).

Discretization of the PEC Adaptive Policy

For each $n = 0,1,\ldots$, let P_n, X_n, $D(n)$, and K_n be as defined in Section 6.3. Thus the discretized θ-model is $\mathrm{CM}_n(\theta) = (X_n, A_n, q_n(\theta), r_n(\theta))$, where, for each $k = (x,a) \in K_n$ and $j = 1,\ldots,m_n$,

$$q_n(x_j^n \mid k,\theta) := q(X_j^n \mid k,\theta) \quad \text{and} \quad r_n(k,\theta) := r(k,\theta).$$

Instead of 3.3, the DPE for $\mathrm{CM}_n(\theta)$ is now

$$\textbf{4.4} \qquad v_n^*(x,\theta) = \max_{a \in A(x)} \left\{ r(x,a,\theta) + \beta \sum_{j=1}^{m_n} v_n^*(x_j^n,\theta)\, q_n(x_j^n \mid x,a,\theta) \right\}$$

for $x \in X_n$, and moreover, there exists an optimal stationary policy $f_n^*(\cdot,\theta)$ for $\mathrm{CM}_n(\theta)$ such that $f_n^*(x,\theta)$ maximizes the r.h.s. of 4.4 for each $x \in X_n$. Finally, we extend $v_n^*(x,\theta)$ and $f_n^*(x,\theta)$ to all $x \in X$ as in 3.4 and 3.5, respectively.

To state the adaptive version of Theorem 3.9, and also for the purpose of comparison with the results in Section 2.5, let us define

4.5 (a) $d(n,\theta) := c^* D(n) + c_1 \cdot \max\{\rho(n,\theta), \pi(n,\theta)\}$, and

(b) $\overline{d}(n,\theta) := 2(1-\beta)^{-1}[c^* D(n) + c_1 \cdot \max\{\overline{\rho}(n,\theta), \overline{\pi}(n,\theta)\}]$,

where $c_1 := (1+\beta c_0)/(1-\beta)$, and c_0 and c^* are the constants in 2.4 and Theorem 3.9, respectively. With this notation, and Assumptions 4.1 and 4.3, we have:

4.6 Theorem. *For any sequence $\{\theta_n\}$ that converges to θ, and $n = 0,1,\ldots$,*

(a) $\|v_n^*(\cdot,\theta_n) - v^*(\cdot,\theta)\| \leq (1-\beta)^{-1}[L^* D(n) + \rho(n,\theta) + \beta c_0 \pi(n,\theta)] \leq d(n,\theta)$;

(b) $\sup_x |\phi(x, f_n^*(x,\theta_n),\theta)| \le 2d(n,\theta)$, *where* $\phi(x,a,\theta)$ *is the discrepancy function in Section* 2.5, *Theorem* 5.2;

(c) *The policy* $\delta := \{f_n^*(\cdot,\theta_n)\}$ *that uses the control* $a_n := f_n^*(x_n,\theta_n)$ *at time* $n = 0,1,\ldots$, *is such that*

$$\sup_x |V_n(\delta,x,\theta) - E_x^{\delta,\theta} v^*(x_n,\theta)| \le \bar{d}(n,\theta), \quad \textit{and}$$

(d) δ *is asymptotically discount optimal (ADO) for the control model* CM(θ).

(e) *Let* $\hat{\theta}_n = \hat{\theta}_n(h_n)$ *be a sequence of strongly consistent estimators (see definition in Section* 2.5 *or Section* 5.1) *of the true parameter value, say* θ^*, *and let* $\delta^* = \{\delta_n^*\}$ *be the policy defined by*

$$\delta_n^*(h_n) := f_n^*(x_n, \hat{\theta}_n(h_n)) \quad \textit{for} \quad h_n \in H_n \quad \textit{and} \quad n = 0,1,\ldots.$$

Then when θ, θ_n, *and* δ *are replaced, respectively, by* θ^*, $\hat{\theta}_n$, *and* δ^*, *the sequences in parts* (a) *and* (b) *converge to zero* $P_x^{\pi^*,\theta^*}$*-almost surely for every* $x \in X$; *part* (c) *holds replacing* $\bar{d}(n,\theta)$ *by the expected value* $E_x^{\pi^*,\theta^*}[\bar{d}(n,\theta^*)]$, *which tends to zero for every* $x \in X$ *as* $n \to \infty$; *finally,* (d) *holds for* δ^*, *i.e.,* δ^* *is ADO for* CM(θ^*).

Notice that, in analogy with the PEC adaptive policy in Section 2.5, δ^* is also obtained by "substituting the estimates, $\hat{\theta}_n$, into the optimal stationary controls" $f_n^*(x,\theta^*)$ for CM$_n(\theta^*)$. We can thus call δ^* a "discretized" version of the PEC policy. Observe also that the r.h.s. of the inequality in Theorem 4.6(a) consists of two parts: a term $c^*D(n)$ corresponding to the discretization, as in Theorem 3.9(a), and a second term $c_1 \cdot \max\{\rho(n,\theta), \pi(n,\theta)\}$ corresponding to the parameter adaptation, as in Proposition 5.6 of Section 2.5.

The proof of Theorem 4.6 is given in the next section. We now turn our attention to the recursive functions 3.10, which yield a "discretization" of the NVI adaptive policy.

Discretization of the NVI Adaptive Policy

Given any sequence $\{\theta_n\}$ in Θ, the parameter-adaptive version of the functions $v_n'(x)$ in 3.10 is,

4.7 $$v_n'(x,\theta_n) := \max_{a\in A(x)} \left\{ r(x,a,\theta_n) + \beta \sum_{j=1}^{m_n} v_{n-1}'(x_j^n, \theta_{n-1}) q_n(x_j^n \mid x,a,\theta_n) \right\}$$

for $x \in X_n$ and $n \ge 0$, where $v_{-1}'(\cdot) \equiv 0$. We then take control functions $f_n'(x,\theta_n) \in A(x)$ maximizing the r.h.s. of 4.7 for $x \in X_n$, and finally, $v_n'(x,\theta_n)$ and $f_n'(x,\theta_n)$ are extended to all $x \in X$ as in Section 6.3.

4.8 Theorem. *For any sequence* $\{\theta_n\}$ *that converges to* θ, *and* $n = 0,1,\ldots$,

(a) $\|v'_n(\cdot,\theta_n) - v^*(\cdot,\theta)\| \leq c_3\overline{I}(n)$, *where* $c_3 := \frac{1}{2}(1-\beta) + 2c_0$, *and*

$$\overline{I}(n) := \max\{\beta^{[n/2]}, \overline{d}([n/2],\theta)\}.$$

(b) $\sup_x |\phi(x, f'_n(x,\theta_n),\theta)| \leq (1+\beta c_0)\max\{\rho(n,\theta),\pi(n,\theta)\} + (1+\beta)c_3\overline{I}(n-1) =: J(n)$.

(c) *The policy* $\delta' = \{f'_n(\cdot,\theta_n), n = 0,1,\ldots\}$ *satisfies*

$$\sup_x |V_n(\delta',x,\theta) - E_x^{\delta',\theta} v^*(x_n,\theta)| \leq (1-\beta)^{-1}\overline{J}(n),$$

where $\overline{J}(n)$ *is the same as* $J(n)$ *with* $\rho(n,\theta)$ *and* $\pi(n,\theta)$ *replaced by* $\overline{\rho}(n,\theta)$ *and* $\overline{\pi}(n,\theta)$, *respectively.*

(d) *The policy* δ' *is asymptotically discount optimal for* CM(θ).

(e) *Theorem* 4.6(e) *holds (in the present NVI context) if* δ^* *is replaced by the adaptive policy* $\hat{\delta} := \{f'_n(\cdot,\hat{\theta}_n)\}$.

The modified-PEC adaptive policy in Section 2.5 can also be discretized in the obvious manner, following the above ideas.

6.5 Proofs

First let us consider the *local income function* $G(k,v)$ defined for all $k = (x,a) \in \mathbf{K}$ by

5.1 $$G(k,v) := r(k) + \beta \int v(y)\, q(dy \mid k).$$

Then using Assumptions 2.1 and inequality B.1 (Appendix B), straightforward calculations yield:

5.2 Lemma. *For any two functions* v *and* v' *in* $B(X)$, *and any two points* k *and* k' *in* $\mathbf{K}$,

(a) $|G(k,v) - G(k,v')| \leq \beta\|v - v'\|$,

(b) $|G(k,v) - G(k',v)| \leq (L_0 + \beta L_2\|v\|)\, d(k,k')$,

(c) $|G(k,v) - G(k',v')| \leq (L_0 + \beta L_2\|v\|)\, d(k,k') + \beta\|v - v'\|$.

The Non-Adaptive Case

Proof of Theorem 3.9. (a) To begin, notice that in terms of the local income function $G(k,v)$ we can write the DPE's 2.2 and 3.6 as

5.3 $$v^*(x) = \max_{a\in A(x)} G(x,a,v^*) \quad \text{for } x \in X,$$

and

5.4 $$v_n^*(x) = \max_{a \in A(x)} G(x, a, v_n^*) \quad \text{for } x \in X_n,$$

respectively. Thus, from Proposition A.3 (Appendix A) and Lemma 5.2(a),

$$\begin{aligned} |v_n^*(x) - v^*(x)| &\leq \max_{a \in A(x)} |G(x, a, v_n^*) - G(x, a, v^*)| \\ &\leq \beta \|v_n^* - v^*\| \quad \text{for all } x \in X_n. \end{aligned}$$

On the other hand, the Lipschitz property 2.3 of v^* implies that, for any $x \in X_i^n$ and $i = 1, \ldots, m_n$,

$$|v^*(x) - v^*(x_i^n)| \leq L^* d_1(x, x_i^n) \leq L^* D(n),$$

and therefore, for any $x \in X_i^n$ and $i = 1, \ldots, m_n$,

$$\begin{aligned} |v_n^*(x) - v^*(x)| &\leq |v_n^*(x_i^n) - v^*(x_i^n)| + |v^*(x_i^n) - v^*(x)| \\ &\leq \beta \|v_n^* - v^*\| + L^* D(n), \end{aligned}$$

from which part (a) follows.

(b) By definition, $f_n^*(x)$ maximizes the r.h.s. of 5.4 for all $x \in X_n$, so that

5.5 $$v_n^*(x) = G(x, f_n^*(x), v_n^*) \quad \text{for all } x \in X_n.$$

Notice also that the discrepancy function $\phi(x, a)$ can be written as

$$\phi(x, a) = G(x, a, v^*) - v^*(x).$$

Thus taking $x \in X_i^n$ and $a = f_n^*(x)$, we obtain,

$$\phi(x, f_n^*(x)) = G(x, f_n^*(x), v^*) - v^*(x).$$

Next, on the r.h.s. add and substract $v_n^*(x) = v_n^*(x_i^n)$ as given in 5.5, and then use Lemma 5.2(c) to get, for all $x \in X_i^n$ and $i = 1, \ldots, m_n$,

$$|\phi(x, f_n^*(x))| \leq (L_0 + \beta L_2 \|v^*\|) \cdot d[(x, f_n^*(x)), (x_i^n, f_n^*(x_i^n))]$$

$$+ (1 + \beta) \|v_n^* - v^*\|$$

[by 3.8, 2.4, and part (a)]

$$\begin{aligned} &\leq L^* D(n) + (1 + \beta) c^* D(n) \\ &= 2c^* D(n), \end{aligned}$$

which proves (b).

(c) This follows from part (b) and equation 3.5 in Section 2.3, using the assumption that $D(n)$ is nonincreasing.

(d) This is a special case of part (c), and finally (e) follows from either (b) or (c), using Theorem 3.6 in Section 2.3. This completes the proof of Theorem 3.9. □

Proof of Theorem 3.12. We begin by noting that 3.10 can also be written as

5.6
$$\begin{aligned} v'_{n+1}(x) &= \max_{a\in A(x)} \left\{ r(x,a) + \beta \int v'_n(y)\, q(dy \mid x,a) \right\} \\ &= \max_{a\in A(x)} G(x,a,v'_n) \quad \text{for } x \in X_{n+1} \text{ and } n \geq 0. \end{aligned}$$

Thus, from 5.3 and Lemma 5.2(a),

$$\begin{aligned} |v'_{n+1}(x) - v^*(x)| &\leq \max_{a\in A(x)} |G(x,a,v'_n) - G(x,a,v^*)| \\ &\leq \beta \|v'_n - v^*\| \quad \text{for } x \in X_{n+1}, \end{aligned}$$

and, on the other hand, the Lipschitz condition 2.3 and the definition of $D(n)$ imply

$$|v^*(x) - v^*(x_i^{n+1})| \leq L^* D(n+1) \quad \text{for all } x \in X_i^{n+1} \text{ and } i = 1, \ldots, m_{n+1}.$$

The last two inequalities together imply

$$\|v'_{n+1} - v^*\| \leq \beta \|v'_n - v^*\| + L^* D(n+1) \quad \text{for all } n \geq 0.$$

Therefore, for all $n \geq 0$ and $m \geq 1$,

$$\begin{aligned} \|v'_{n+m} - v^*\| &\leq \beta^m \|v'_n - v^*\| + L^* \sum_{j=0}^{m-1} \beta^j D(n+m-j) \\ &\leq 2c_0\beta^m + (1-\beta)^{-1} L^* D(n) \end{aligned}$$

[since $\|v'_n - v^*\| \leq 2c_0$ for all $n \geq 0$, and $D(n)$ is nonincreasing]

$$\leq (2c_0 + c^*) \cdot \max\{\beta^m, D(n)\}.$$

Finally, if we let $t = n + m$, $n = [t/2]$, and $m = t - n \geq [t/2]$, the above inequality becomes

$$\|v'_t - v^*\| \leq c_2 \cdot \max\{\beta^{[t/2]}, D([t/2]\}$$

which proves part (a).

(b) This proof is very similar to that of Theorem 3.9(b). Namely, we first write the discrepancy function as

$$\phi(x, f'_n(x)) = G(x, f'_n(x), v^*) - v^*(x) \quad \text{for } x \in X_i^n \text{ and } i = 1, \ldots, m_n.$$

Next on the r.h.s. we add and subtract

$$v'_n(x) = v'_n(x_i^n) = G(x_i^n, f'_n(x_i^n), v'_{n-1}) \quad \text{for } x \in X_i^n,$$

and then a direct calculation using Lemma 5.2(c) and inequality 3.8, with f_n^* replaced by f_n', yields

$$|\phi(x, f_n'(x))| \leq L^* D(n) + \|v_n' - v^*\| + \beta\|v_{n-1}' - v^*\|.$$

This proves the first inequality in (b), and the second follows from part (a).

Parts (c) and (d) are concluded exactly as parts (c) and (e), respectively, in the proof of Theorem 3.9. □

The Adaptive Case

The proofs in this case are very similar to the non-adaptive case, and therefore, we will sketch the main ideas and leave the details to the reader.

As usual we begin with the changes in notation. For instance, the local income function G in 5.1 is now written as

5.7 $$G(k, v, \theta) := r(k, \theta) + \beta \int v(y)\, q(dy \,|\, k, \theta),$$

where the function v may depend on θ, $v(y, \theta)$, and the θ-DPE in 4.2 becomes (cf. 5.3)

5.8 $$v^*(x, \theta) = \max_{a \in A(x)} G(x, a, v^*(\cdot, \theta), \theta) \quad \text{for } x \in X.$$

Similarly, the θ-version of Lemma 5.2 is the following.

5.9 Lemma. *Parts* (a), (b) *and* (c) *are exactly as in Lemma* 5.2 *with the change*

$$G(k, v) \to G(k, v, \theta) \quad \textit{for all} \quad k \in \mathbf{K} \quad \textit{and} \quad v \in B(X).$$

(d) $|G(k, v, \theta) - G(k, v, \theta')| \leq |r(k, \theta) - r(k, \theta')| + \beta\|v\| \, \|q(\cdot \,|\, k, \theta) - q(\cdot \,|\, k, \theta')\|$; *in particular, for any sequence* θ_n *and* θ *as in Assumptions* 4.3,

(e) $|G(k, v, \theta_n) - G(k, v, \theta)| \leq \rho(n, \theta) + \beta\|v\| \, \pi(n, \theta)$.

Proof of Theorem 4.6. (a) First, we write the DPE 4.4 as (cf. 5.4)

5.10 $$v_n^*(x, \theta) = \max_{a \in A(x)} G(x, a, v_n^*(\cdot, \theta), \theta) \quad \text{for } x \in X_n.$$

Replacing θ by θ_n, comparison with 5.8 yields: for any $x \in X_n$, and writing k for (x, a),

$$|v_n^*(x, \theta_n) - v^*(x, \theta)| \leq \max_{a \in A(x)} |G(k, v_n^*(\cdot, \theta_n), \theta_n) - G(k, v^*(\cdot, \theta), \theta)|.$$

Next, use Lemma 5.9, the Lipschitz condition of the optimal reward function $x \to v^*(x, \theta)$, which is the same as 2.3, and the fact that $\|v_n^*(\cdot, \theta_n)\| \leq c_0$ for all n, to obtain

$$|v_n^*(x, \theta_n) - v^*(x, \theta)| \leq \beta\|v_n^*(\cdot, \theta_n) - v^*(\cdot, \theta)\| + \rho(n, \theta)$$

$$+ \beta c_0 \pi(n,\theta) + L^* D(n) \quad \text{for all} \quad x \in X,$$

which implies part (a).

(b) To simplify the notation, we write $k_n := (x, f_n^*(x,\theta_n))$, so that the discrepancy function we want to estimate becomes

$$\textbf{5.11} \qquad \phi(x, f_n^*(x,\theta_n),\theta) = \phi(k_n,\theta) = G[k_n, v^*(\cdot,\theta),\theta] - v^*(x,\theta).$$

In addition, for $i = 1, \ldots, m_n$, write $(x_i^n, f_n^*(x_i^n,\theta_n))$ as k_i^n. Next, for any $i = 1, \ldots, m_n$ and $x \in X_i^n$, on the r.h.s. of 5.11 add and subtract each of the following terms: $G[k_i^n, v^*(\cdot,\theta),\theta]$, and

$$v_n^*(x,\theta_n) = v_n^*(x_i^n,\theta_n) = G[k_i^n, v_n^*(\cdot,\theta_n),\theta_n].$$

Then a straightforward calculation using Lemma 5.9 and part (a) of the theorem yields part (b).

(c) First note that $2d(n,\theta) \leq (1-\beta)\overline{d}(n,\theta)$. Now using equation 3.5 in Section 2.3, and part (b), we see that

$$\sup_x |V_n(\delta,x,\theta) - E_x^{\delta,\theta} v^*(x_n,\theta)| \leq \sum_{t=n}^{\infty} \beta^{t-n} 2d(t,\theta) \leq \overline{d}(n,\theta),$$

since $\overline{d}(n,\theta)$ is nonincreasing.

Finally, (d) follows from either (b) or (c) and the results in Section 2.3, whereas part (e) is obtained from the strong consistency of $\hat{\theta}_n$ and the arbitrariness of θ and θ_n in parts (a)–(d). □

Proof of Theorem 4.8. (a) Let us rewrite 4.7 as

$$\textbf{5.12} \qquad v'_{n+1}(x,\theta_{n+1}) = \max_{a \in A(x)} G(x,a,v'_n(\cdot,\theta_n),\theta_{n+1}) \quad \text{for} \quad x \in X_{n+1}.$$

Thus, from the DPE 5.8 and Lemma 5.9,

$$\begin{aligned} |v'_{n+1}(x,\theta_{n+1}) - v^*(x,\theta)| &\leq \rho(n+1,\theta) + \beta c_0 \pi(n+1,\theta) \\ &\quad + \beta \|v'_n(\cdot,\theta_n) - v^*(\cdot,\theta)\|. \end{aligned}$$

Next use the Lipschitz condition 2.3 for $x \to v^*(x,\theta)$ to obtain

$$\begin{aligned} \|v'_{n+1}(\cdot,\theta_{n+1}) - v^*(\cdot,\theta)\| &\leq L^* D(n) + (1+\beta c_0)\max\{\rho(n+1,\theta), \pi(n+1,\theta)\} \\ &+ \beta \|v'_n(\cdot,\theta_n) - v^*(\cdot,\theta)\| = (1-\beta)d(n+1,\theta) + \beta\|v'_n(\cdot,\theta_n) - v^*(\cdot,\theta)\|. \end{aligned}$$

Using the inequality $2d(n,\theta) \leq (1-\beta)\overline{d}(n,\theta)$ again, and the fact that $\overline{d}(n,\theta)$ is nonincreasing, we have

$$\begin{aligned} \|v'_{n+m}(\cdot,\theta_{n+m}) - v^*(\cdot,\theta)\| &\leq \frac{1}{2}(1-\beta)^2 \sum_{j=0}^{m-1} \beta^j \overline{d}(n+m-j,\theta) \\ &\quad + \beta^m \|v'_n(\cdot,\theta_n) - v^*(\cdot,\theta)\| \\ &\leq \frac{1}{2}(1-\beta)\overline{d}(n,\theta) + 2c_0\beta^m \\ &\leq c_3 \cdot \max\{\overline{d}(n,\theta), \beta^m\}. \end{aligned}$$

We then conclude part (a) as in the proof of Theorem 3.12(a).

(b) On the r.h.s. of

$$\phi(x, f'_n(x,\theta_n),\theta) = G(x, f'_n(x,\theta_n), v^*(\cdot,\theta),\theta) - v^*(x,\theta)$$

add and subtract $v'_n(x,\theta_n)$ as given in 5.12, and also add and subtract

$$G(x, f'_n(x,\theta_n), v^*(\cdot,\theta),\theta_n),$$

to obtain, from Lemma 5.9,

$$\begin{aligned}\sup_x |\phi(x, f'_n(x,\theta_n),\theta)| &\le \rho(n,\theta) + \beta\|v^*\|\,\pi(n,\theta) + \|v'_n(\cdot,\theta_n) - v^*(\cdot,\theta)\| \\ &+ \ \beta\|v'_{n-1}(\cdot,\theta_{n-1}) - v^*(\cdot,\theta)\| \\ &\le \ (1+\beta c_0)\cdot\max\{\rho(n,\theta),\pi(n,\theta)\} + (1+\beta)c_3\bar{I}(n-1),\end{aligned}$$

since $\bar{I}$ is nonincreasing.

Parts (c), (d) and (e) can be concluded as in the proof of Theorem 4.6. □

6.6 Comments and References

Most of the material in this chapter is contained in Hernández-Lerma and Marcus (1989a), where the reader can also find discretizations for the non-parametric control model $x_{t+1} = F(x_t, a_t, \xi_t)$ in Section 2.6. In the same paper it is also remarked that the approximation to dynamic programming (DP) problems by state discretizations was pioneered by Bellman and Dreyfus (1962) and further developed by other authors. In particular, the non-recursive (non-adaptive) discretization in Section 6.3 is essentially the one introduced by Bertsekas (1975).

There are many other ways to approximate DP problems; see, e.g., Di-Masi and Runggaldier (1986), Haurie and L'Ecuyer (1986), Hinderer and Hübner (1977), in addition to several papers in Puterman (1978) and the references cited for Sections 2.4 and 3.4–3.6. Of particular importance are the works by Langen (1981) and Whitt (1978/79) who analyze approximation procedures of great generality. The reason why we have chosen the discretize-the-state approach is that, as shown in Sections 6.3 and 6.4, it is easily made recursive and it can be extended in a natural way to adaptive control problems. Related works for *denumerable* state adaptive problems are those of Cavazos-Cadena (1987) and Acosta Abreu (1987).

Appendix A

Contraction Operators

Let (V, d) be a metric space. A function T from V into itself is said to be a *contraction operator* if for some β satisfying $0 \leq \beta < 1$ (and called the *modulus* of T) one has

$$d(Tu, Tv) \leq \beta d(u, v) \quad \text{for all } \; u \text{ and } v \text{ in } V.$$

An element v^* of V is called a *fixed point* of T if $Tv^* = v^*$. For a function $T : V \to V$, the function T^n is defined recursively by $T^n := T(T^{n-1})$ for all $n = 1, 2, \ldots$, where T^0 is the identity function.

A.1 Proposition (Banach's Fixed Point Theorem). *If T is a contraction operator mapping a complete metric space (V, d) into itself, then T has a unique fixed point, say, v^*. Furthermore, for any $v \in V$ and $n \geq 0$,*

$$d(T^n v, v^*) \leq \beta^n d(v, v^*).$$

The function $v_n := Tv_{n-1} = T^n v$ are called the successive approximations.

Proof. Ross (1970), Luenberger (1969), ... □

In the text, the metric space V is usually $B(X)$, the Banach space of real-valued, bounded, measurable functions on a Borel space X with the sup norm, $||v|| := \sup_x |v(x)|$. In such a case, we have the following.

A.2 Proposition. *Let T be an operator from $B(X)$ into itself and suppose that:*

(a) *T is monotone, i.e., if u and v are in $B(X)$ and $u \leq v$, then $Tu \leq Tv$.*

(b) *There is a number $\beta \in [0, 1)$ such that $T(v + c) = Tv + \beta c$ for all $v \in B(X)$ and any constant c.*

Then T is a contraction operator with modulus β.

Indeed, applying T to $u(\cdot) \leq v(\cdot) + ||u - v||$, conditions (a) and (b) imply $Tu(x) \leq Tv(x) + \beta||u - v||$, or

$$Tu(x) - Tv(x) \leq \beta||u - v|| \quad \text{for all } \; x \in X.$$

Interchanging u and v and combining the result with the latter inequality, we obtain

$$|Tu(x) - Tv(x)| \leq \beta||u - v|| \quad \text{for all } \; x \in X,$$

which implies the desired result: $\|Tu - Tv\| \leq \beta\|u - v\|$ for u and v in $B(X)$.

Another useful result to check (e.g.) the contraction property for operators defined by a maximization operation is the following.

A.3 Proposition. *Let X be an arbitrary non-empty set, and let u and v be functions from X to* $\mathbf{R}$ *bounded from above (so that* $\sup u$ *and* $\sup v$ *are finite). Then*

$$|\sup_x u(x) - \sup_x v(x)| \leq \sup_x |u(x) - v(x)|.$$

Proof. Hinderer (1970), p. 17. □

Appendix B

Probability Measures

Total Variation Norm

Let $(X, \mathcal{B})$ be a measurable space. An extended real-valued function μ defined on the sigma-algebra $\mathcal{B}$ is called a ***signed measure*** if $\mu(\emptyset) = 0$, μ is sigma-additive, and μ attains at most one of the values $+\infty$, $-\infty$. [If $\mu(B) \geq 0$ for all $B \in \mathcal{B}$, then, as usual, μ is simply called a ***measure***.]

For a signed measure μ we define its positive and negative parts by

$$\mu^+(B) := \sup\{\mu(A) \mid A \in \mathcal{B} \text{ and } A \subset B\}$$

and

$$\mu^-(B) := -\inf\{\mu(A) \mid A \in \mathcal{B} \text{ and } A \subset B\}.$$

These are measures on X, at least one of them is finite, and $\mu = \mu^+ - \mu^-$. The latter equality is called the Hahn (or Jordan–Hahn) decomposition of μ (see also B.0 below) and

$$\|\mu\| := \mu^+(X) + \mu^-(X) = \sup_B \mu(B) - \inf_B \mu(B) \quad (\text{where } B \in \mathcal{B})$$

is the (*total*) *variation* of μ [Ash (1972), Halmos (1950), Royden (1968),...]. $\|\cdot\|$ is a norm on the vector space of all *finite* signed measures on X.

B.0 Jordan–Hahn Decomposition Theorem. *If μ is a signed measure on $(X, \mathcal{B})$, then there exist disjoint measurable sets X^+ and X^- whose union is X and such that*

$$\mu(X^+ \cap B) \geq 0 \quad \textit{and} \quad \mu(X^- \cap B) \leq 0 \quad \textit{for all} \quad B \in \mathcal{B},$$

and, moreover, the variation norm of μ is given by

$$\|\mu\| = \mu(X^+) - \mu(X^-).$$

Proof. Ash (1972), pp. 60–61; Royden (1968), pp. 235–236; .... □

Integration with respect to a (finite) signed measure μ is defined by

$$\int v \, d\mu := \int v \, d\mu^+ - \int v \, d\mu^-.$$

If $v \in B(X)$, then

B.1
$$\left| \int v \, d\mu \right| \leq \|v\| \, \|\mu\|.$$

Let P and Q be probability measures on $(X, \mathcal{B})$. Then $P - Q$ is a finite signed measure and its total variation is given by

B.2
$$\|P - Q\| = 2 \sup_{B \in \mathcal{B}} |P(B) - Q(B)|.$$

If P and Q have densities p and q with respect to some sigma-finite measure λ on X, then by Scheffé's Theorem [see, e.g., Devroye and Györfi (1985), p. 2],

B.3
$$\|P - Q\| = \int |p - q| \, d\lambda,$$

and if $P = \{p(i)\}$ and $Q = \{q(i)\}$ are *discrete* probability distributions, then
$$\|P - Q\| = \sum_i |p(i) - q(i)|.$$

Another result, also called Scheffé's Theorem, is the following. Suppose that P_n and P are probability measures with densities p and p_n with respect to a sigma-finite measure λ on X, where $n = 1, 2, \ldots$, and such that $p_n(x) \to p(x)$ for λ-almost all x. Then

$$\|P_n - P\| = \int |p_n(x) - p(x)| \, \lambda(dx) \to 0.$$

Moreover, by a result of Glick (1974), the latter convergence also holds a.s., if the densities $p_n(x) = p_n(x, \omega)$ are random and, as $n \to \infty$,

$$p_n(x) \to p(x) \quad \text{a.s. for } \lambda\text{-almost all } x.$$

A related result is the following.

B.4 Proposition. [Royden (1968), p. 232]. *Let $\{\mu_n\}$ be a sequence of measures on $(X, \mathcal{B})$ which converges setwise to a measure μ, that is, $\mu_n(B) \to \mu(B)$ for all $B \in \mathcal{B}$. Let $\{f_n\}$ and $\{g_n\}$ be two sequences of measurable functions converging pointwise to f and g, respectively, and suppose that $|f_n| \leq g_n$ and*

$$\lim \int g_n d\mu_n = \int g \, d\mu < \infty.$$

Then
$$\lim \int f_n d\mu_n = \int f \, d\mu.$$

Weak Convergence

Throughout the following, X is assumed to be a Borel space, and $\mathcal{B} = \mathcal{B}(X)$ is the Borel sigma-algebra. We denote by $\mathbf{P}(X)$ the space of probability measures on X. (Sometimes we use the abbreviation p.m. for "probability measure".)

Let $P, P_1, P_2, \ldots$ be p.m.'s on X. It is said that P_n *converges weakly* to P if, as $n \to \infty$,

B.5
$$\int f\, dP_n \to \int f\, dP \quad \text{for all} \quad f \in C(X),$$

where $C(X)$ is the space of real-valued, bounded, continuous functions on X with the sup norm. Equivalently [Ash (1972), Bertsekas and Shreve (1978), Parthasarathy (1967),...], P_n converges weakly to P if

$$\int u\, dP_n \to \int u\, dP$$

for every uniformly continuous function u in $C(X)$. There are several other equivalent statements of weak convergence (see, e.g., the references above).

We will always understand $\mathbf{P}(X)$ as a topological space with the topology of weak convergence. In such a case, since X is a Borel space, $\mathbf{P}(X)$ is also a Borel space. For each Borel subset B of X, let $g_B : \mathbf{P}(X) \to [0,1]$ be the mapping defined by

B.6
$$g_B(P) := P(B).$$

Then the Borel sigma-algebra of $\mathbf{P}(X)$ is the smallest sigma-algebra with respect to which g_B is measurable for every $B \in \mathcal{B}(X)$; see, e.g., Bertsekas and Shreve (1978), p. 133.

B.7 Definition. Let P be a p.m. on X, and let $\mathcal{H}$ be a family of real-valued measurable functions on X. It is said that $\mathcal{H}$ is a *P-uniformity class* if

$$\sup_{h \in \mathcal{H}} \left| \int h\, dP_n - \int h\, dP \right| \to 0 \quad \text{as} \quad n \to \infty,$$

for any sequence $\{P_n\}$ in $\mathbf{P}(X)$ which converges weakly to P.

B.8 Proposition. [Billingsley and Topsoe (1967); Billingsley (1968), p. 17]. *Let $\mathcal{H}$ be a family of real-valued functions on X. If $\mathcal{H}$ is uniformly bounded and equicontinuous at each $x \in X$, then $\mathcal{H}$ is a P-uniformity class for every* p.m. *on X. ($\mathcal{H}$ equicontinuous at each $x \in X$ means that for each $x \in X$ and $\epsilon > 0$, there exists $\delta > 0$ such that if*

$$d(x,y) < \delta, \ \textit{then} \ |h(x) - h(y)| < \epsilon \ \textit{for all} \ h \in \mathcal{H},$$

where d is the metric on X.)

Appendix C

Stochastic Kernels

An excellent exposition on stochastic kernels is given by Bertsekas and Shreve (1978), Section 7.4.3; most of the following concepts and results are borrowed from that reference.

Let X and Y be Borel spaces. A (Borel-measurable) *stochastic kernel*—or conditional probability measure—on X given Y is a function $q(dx \mid y)$ such that for each $y \in Y$, $q(\cdot \mid y)$ is a probability measure on X, and for each Borel set $B \in \mathcal{B}(X)$, $q(B \mid \cdot)$ is a measurable function from X to $[0,1]$. Equivalently, a collection of probability measures $q(dx \mid y)$ on X parameterized by $y \in Y$ is a stochastic kernel if and only if the function $h : Y \to \mathbf{P}(X)$ defined by

C.1
$$h(y) := q(\cdot \mid y)$$

is measurable.

A stochastic kernel $q(dx \mid y)$ on X given Y is said to be *continuous* if the function h in C.1 is continuous, that is, $q(\cdot \mid y_n)$ converges weakly to $q(\cdot \mid y)$ whenever y_n converges to y. Thus, by B.5 above, the stochastic kernel $q(dx \mid y)$ is continuous if $\int v(x) q(dx \mid y)$ is a continuous function of $y \in Y$ for every function $v \in C(X)$.

C.2 Proposition. *Let $q(dx \mid y)$ be a stochastic kernel on X given Y; let $f(x,y)$ be a real-valued measurable function on XY, and let $f' : Y \to \mathbf{R}$ be the function defined by*

$$f'(y) := \int f(x,y)\, q(dx \mid y)$$

whenever the integral exists.

(a) *If $f \in B(XY)$, then $f' \in B(Y)$.*

(b) *If $q(dx \mid y)$ is continuous and $f \in C(XY)$, then $f' \in C(Y)$.*

The following result is due to C. Ionescu Tulcea, for a proof, see (e.g.) Bertsekas and Shreve, p. 140, or Ash (1972), Section 2.7.

C.3 Proposition. *Let $X_1, X_2, \ldots$ be a sequence of Borel spaces, and define $Y_n := X_1 X_2 \ldots X_n$, and $Y := X_1 X_2 \ldots$. Let $p \in \mathbf{P}(X_1)$ be a given p.m., and for $n = 1, 2, \ldots$, let $q_n(dx_{n+1} \mid y_n)$ be a stochastic kernel on X_{n+1} given*

Y_n. Then for each $n \geq 2$, there exists a unique p.m. $r_n \in \mathbf{P}(Y_n)$ such that for all $B_i \in \mathcal{B}(X_i)$, where $i = 1, \ldots, n$,

$$r_n(B_1 B_2 \ldots B_n) = \int_{B_1} p(dx_1) \int_{B_2} q_1(dx_2 \mid x_1) \cdots$$

$$\cdot \int_{B_n} q_{n-1}(dx_n \mid x_1, \ldots, x_{n-1}).$$

(In the text, sometimes we write r_n as

$$r_n(dx_1 dx_2 \ldots dx_n) = p(dx_1) q_1(dx_2 \mid x_1) \cdots q_{n-1}(dx_n \mid x_1, \ldots, x_{n-1}),$$

or we use the shorter notation $r_n = pq_1q_2 \ldots q_{n-1}$.) Moreover, if a measurable function f on Y_n is r_n-integrable, then

$$\int_{Y_n} f \, dr_n = \int_{X_1} p(dx_1) \int_{X_2} q_1(dx_2 \mid x_1) \cdots$$

$$\cdot \int_{X_n} f(x_1, \ldots, x_n) q_{n-1}(dx_n \mid x_1, \ldots, x_{n-1}).$$

Finally, there exists a unique p.m. r on $Y = X_1 X_2 \ldots$, sometimes written as

$$r = pq_1q_2 \ldots,$$

such that, for each n, the marginal of r on Y_n is r_n.

Appendix D

Multifunctions and Measurable Selectors

The Hausdorff Metric

Let (A, d) be a metric space. If B is a subset of A and $a \in A$, we define

$$d(a, B) := \inf\{d(a, b) \mid b \in B\}.$$

If B_1 and B_2 are two non-empty closed subsets of A, we write

$$d(B_1, B_2) := \sup\{d(b_1, B_2) \mid b_1 \in B_1\}$$

and

$$d(B_2, B_1) := \sup\{d(b_2, B_1) \mid b_2 \in B_2\}.$$

The function

$$H(B_1, B_2) := \max\{d(B_1, B_2), d(B_2, B_1)\}$$

is called the Hausdorff metric and it satisfies the properties of a metric on the family of non-empty closed subsets of A.

Let $C(A)$ be the collection of all non-empty compact subsets of A topologized by the Hausdorff metric H. If (A, d) is separable, then $(C(A), H)$ is a separable metric space [Berge (1963); Schäl (1975)].

Multifunctions

Throughout the following, X and A denote Borel spaces. A mapping D which associates with each $x \in X$ a non-empty subset $D(x)$ of A is called a *multifunction* (or correspondence, or set-valued function) from X to A. The set

$$\mathbf{K} := \{(x, a) \mid x \in X \text{ and } a \in D(x)\}$$

is called the graph of D.

In the text, X and A denote, respectively, the state space and the action (or control) set in a MCM (X, A, q, r), and $D(x) = A(x)$ is the set of admissible actions in state $x \in X$; see Section 1.2.

A multifunction D from X to $C(A)$ is said to be *upper semicontinuous* (u.s.c.) if for each open subset A' of A the set $\{x \mid D(x) \text{ is contained in } A'\}$ is open in X. Equivalently, D is u.s.c. if and only if for any $x \in X$ and any open set A' containing $D(x)$ there is a neighborhood N of x such that $D(N)$ is contained in A'. [Schäl (1975).]

D.1 Definition. A multifunction D from X to A is said to be *Borel-measurable* if the set

$$D^{-1}[B] := \{x \in X \mid D(x) \cap B \neq \emptyset\}$$

is a Borel subset of X for every closed subset B of A, and a (measurable) *selector* or *decision function* for D is a measurable function $f : X \to A$ such that $f(x) \in D(x)$ for all x in X. (In the text, the set of all selectors is denoted by $\mathbf{F}$, and is identified with the set of stationary policies; see Definition 2.4 in Chapter 1.)

D.2 Proposition. *Let D be a multifunction from X to A with compact values, that is, $D : X \to \mathcal{C}(A)$. Then the following statements are equivalent:*

(a) *D is Borel-measurable.*

(b) *$D^{-1}[B]$ is a Borel set in X for every open subset B of A.*

(c) *The graph $\mathbf{K}$ of D is a Borel subset of (the product space) XA.*

(d) *D is a Borel-measurable function from X to the space $\mathcal{C}(A)$ (topologized by the Hausdorff metric).*

Proof. Himmelberg et al. (1976). □

Recall that a real-valued function v on X is said to be *upper semicontinuous* = u.s.c. [respectively, *lower semicontinuous* = l.s.c.] if for any sequence $\{x_n\}$ in X converging to $x \in X$, we have $\limsup v(x_n) \leq v(x)$ [respectively, $\liminf v(x_n) \geq v(x)$]. Clearly, v is continuous if and only if v is both u.s.c. and l.s.c.

Proposition D.3(a) below is called a *Measurable Selection Theorem.*

D.3 Proposition. *Let $D : X \to \mathcal{C}(A)$ be a Borel-measurable multifunction, and let $v(x, a)$ be a real-valued measurable function on $\mathbf{K}$ such that $v(x, a)$ is upper semicontinuous (u.s.c.) in $a \in D(x)$ for each $x \in X$. Then:*

(a) *There exists a selector $f : X \to A$ for D such that*

$$v(x, f(x)) = \max_{a \in D(x)} v(x, a) \quad \text{for every } x \in X,$$

and the function $v^*(x) := \max_{a \in D(x)} v(x, a)$ is measurable.

(b) *If D is u.s.c. and v is u.s.c. and bounded (from above), then v^* is u.s.c. and bounded (respectively, from above).*

(c) *If D is continuous and v is continuous and bounded, then v^* is continuous and bounded.*

(d) *If D is continuous and A is compact, then* $\mathbf{K}$ *is closed.*

Proof. Himmelberg et al. (1976); Schäl (1975); Bertsekas and Shreve (1978). □

References

Acosta Abreu, R.S. (1987). Adaptive Markov decision processes with average reward criterion, Doctoral Thesis. Departamento de Matemáticas, CINVESTAV-IPN. (In Spanish).

Acosta Abreu, R.S. (1987a). Control de cadenas de Markov con parámetros desconocidos y espacio de estados métrico. Submitted for publication.

Acosta Abreu, R.S. (1987b). Control de procesos de Markov parcialmente observables y con parámetros desconocidos. Submitted for publication.

Acosta Abreu, R.S., and Hernández-Lerma, O. (1985). Iterative adaptive control of denumerable state average-cost Markov systems, *Control and Cyber.* **14**, 313–322.

Andreatta, G., and Runggaldier, W.J. (1986). An approximation scheme for stochastic dynamic optimization problems, *Math. Programm. Study* **27**, 118–132.

Arstein, Z. (1978). Relaxed controls and the dynamics of control systems, *SIAM J. Control Optim.* **16**, 689–701.

Ash, R.B. (1972). *Real Analysis and Probability,* Academic Press, New York.

Baranov, V.V. (1981). Recursive algorithms of adaptive control in stochastic systems, *Cybernetics* **17**, 815–824.

Baranov, V.V. (1982). A recursive algorithm in markovian decision processes, *Cybernetics* **18**, 499–506.

Baum, L.E., Petrie, T., Soules, G., and Weiss, N. (1970). A maximization technique occurring in the statistical analysis of probabilistic functions of Markov chains, *Ann. Math. Statist.* **41**, 164–171.

Bellman, R. (1957). *Dynamic Programming,* Princeton University Press, Princeton, N.J.

Bellman, R. (1961). *Adaptive Control Processes: A Guided Tour,* Princeton University Press, Princeton, N.J.

Bellman, R., and Dreyfus, S.E. (1962). *Applied Dynamic Programming,* Princeton University Press, Princeton, N.J.

Bensoussan, A. (1982). Stochastic control in discrete time and applications to the theory of production, *Math. Programm. Study* **18**, 43–60.

Berge, C. (1963). *Topological Spaces,* Macmillan, New York.

Bertsekas, D.P. (1975). Convergence of discretization procedures in dynamic programming, *IEEE Trans. Autom. Control* **20**, 415–419.

Bertsekas, D.P. (1976). *Dynamic Programming and Stochastic Control,* Academic Press, New York.

Bertsekas, D.P. (1987). *Dynamic Programming: Deterministic and Stochastic Models,* Prentice-Hall, Englewood Cliffs, N.J.

Bertsekas, D.P., and Shreve, S.E. (1978). *Stochastic Optimal Control: The Discrete Time Case,* Academic Press, New York.

Billingsley, P. (1961). *Statistical Inference for Markov Processes,* University of Chicago Press, Chicago.

Billingsley, P. (1968). *Convergence of Probability Measures,* Wiley, New York.

Billingsley, P., and Topsoe, F. (1967). Uniformity in weak convergence, *Z. Wahrsch. Verw. Geb.* **7**, 1–16.

Blackwell, D. (1965). Discounted dynamic programming, *Ann. Math. Statist.* **36**, 226–235.

Borkar, V., and Varaiya, P. (1982). Identification and adaptive control of Markov chains, *SIAM J. Control Optim.* **20**, 470–489.

Cavazos-Cadena, R. (1986). Finite-state approximations for denumerable state discounted Markov decision processes, *Appl. Math. Optim.* **14**, 1–26.

Cavazos-Cadena, R. (1987). Finite-state approximations and adaptive control of discounted Markov decision processes with unbounded rewards, *Control and Cyber.* **16**, 31–58.

Cavazos-Cadena, R. (1988). Necessary conditions for the optimality equation in average-reward Markov decision processes, *Appl. Math. Optim.* (to appear).

Cavazos-Cadena, R. (1988a). Necessary and sufficient conditions for a bounded solution to the optimality equation in average reward Markov decision chains, *Syst. Control Lett.* **10**, 71–78.

Cavazos-Cadena, R. (1988b). Weak conditions for the existence of optimal stationary policies in average Markov decision chains with unbounded costs. Submitted for publication.

Cavazos-Cadena, R., and Hernández-Lerma, O. (1987). Adaptive policies for priority assignment in discrete-time queues—discounted cost criterion. Submitted for publication.

Cavazos-Cadena, R., and Hernández-Lerma, O. (1989). Recursive adaptive control of average Markov decision processes, Reporte Interno. Depto. de Matemáticas, CINVESTAV-IPN.

Clark, C.W. (1976). *Mathematical Bioeconomics: The Optimal Management of Renewable Resources,* Wiley, New York.

Collomb, G. (1981). Estimation non-paramétrique de la régression: revue bibliographique, *Internat. Statist. Rev.* **49**, 75–93.

DeGroot, M.H. (1970). *Optimal Statistical Decisions,* McGraw-Hill, New York.

Devroye, L., and Györfi, L. (1985). *Nonparametric Density Estimation: The L_1 View,* Wiley, New York.

DiMasi, G.B., and Runggaldier, W.J. (1986). An approach to discrete-time stochastic control problems under partial observations, *SIAM J. Control Optim.* **25**, 38–48.

Doob, J.L. (1953). *Stochastic Processes,* Wiley, New York.

Doshi, B.T., and Shreve, S.E. (1980). Strong consistency of a modified maximum likelihood estimator for controlled Markov chains, *J. Appl. Probab.* **17**, 726–734.

Doukhan, P., et Ghindès, M. (1983). Estimation de la transition de probabilité d'une chaine de Markov Doëblin-recurrente, *Stoch. Proc. Appl.* **15**, 271–293.

Dynkin, E.B., and Yushkevich, A.A. (1979). *Controlled Markov Processes,* Springer-Verlag, New York.

El-Fattah, Y.M. (1981). Gradient approach for recursive estimation and control in finite Markov chains, *Adv. Appl. Probab.* **13**, 778–803.

Federgruen, A., and Schweitzer, P.J. (1981). Nonstationary Markov decision problems with converging parameters, *J. Optim. Theory Appl.* **34**, 207–241.

Federgruen, A., and Tijms, H.C. (1978). The optimality equation in average cost denumerable state semi-Markov decision problems, recurrency conditions and algorithms, *J. Appl. Probab.* **15**, 356–373.

Flynn, J. (1976). Conditions for the equivalence of optimality criteria in dynamic programming, *Ann. Statist.* **4**, 936–953.

Gaenssler, P., and Stute, W. (1979). Empirical processes: a survey for i.i.d. random variables, *Ann. Probab.* **7**, 193–243.

Gänssler, P. (1972). Note on minimum contrast estimates for Markov processes, *Metrika* **19**, 115–130.

Georgin, J.P. (1978a). Controle de chaines de Markov sur des espaces arbitraires, *Ann. Inst. H. Poincaré* **14**, Sect. B, 255–277.

Georgin, J.P. (1978b). Estimation et controle des chaines de Markov sur des espaces arbitraires, *Lecture Notes Math.* **636**, Springer-Verlag, 71–113.

Getz, W.M., and Swartzman, G.L. (1981). A probability transition matrix model for yield estimation in fisheries with highly variable recruitment, *Can. J. Fish. Aquat. Sci.* **38**, 847–855.

Glick, N. (1974). Consistency conditions for probability estimators and integrals of density estimators, *Utilitas Math.* **6**, 61–74.

Gordienko, E.I. (1985). Adaptive strategies for certain classes of controlled Markov processes, *Theory Probab. Appl.* **29**, 504–518.

Gubenko, L.G., and Statland, E.S. (1975). On controlled, discrete-time Markov decision processes, *Theory Probab. Math. Statist.* **7**, 47–61.

Hall, P., and Heyde, C.C. (1980). *Martingale Limit Theory and Its Application,* Academic Press, New York.

Halmos, P.R. (1950). *Measure Theory,* Van Nostrand, Princeton, N.J.

Haurie, A., and L'Ecuyer, P. (1986). Approximation and bounds in discrete event dynamic programming, *IEEE Trans. Autom. Control* **31**, 227–235.

Hernández-Lerma, O. (1985). Nonstationary value-iteration and adaptive control of discounted semi-Markov processes, *J. Math. Anal. Appl.* **112**, 435–445.

Hernández-Lerma, O. (1985a). Approximation and adaptive policies in discounted dynamic programming, *Bol. Soc. Mat. Mexicana* **30**, 25–35.

Hernández-Lerma, O. (1986). Finite-state approximations for denumerable multidimensional state discounted Markov decision processes, *J. Math. Anal. Appl.* **113**, 382–389.

Hernández-Lerma, O. (1987). Approximation and adaptive control of Markov processes: average reward criterion, *Kybernetika* (Prague) **23**, 265–288.

Hernández-Lerma, O. (1988). Controlled Markov processes: recent results on approximation and adaptive control, Texas Tech University Math. Series, **15**, 91–117.

Hernández-Lerma, O., and Cavazos-Cadena, R. (1988). Continuous dependence of stochastic control models on the noise distribution, *Appl. Math. Optim.* **17**, 79–89.

Hernández-Lerma, O., and Doukhan, P. (1988). Nonparametric estimation and adaptive control of a class of discrete-time stochastic systems. Submitted for publication.

Hernández-Lerma, O., and Duran, B.S. (1988). Nonparametric density estimation and adaptive control. Submitted for publication.

Hernández-Lerma, O., Esparza, S.O., and Duran, B.S. (1988). Recursive nonparametric estimation of nonstationary Markov processes. Submitted for publication.

Hernández-Lerma, O., and Lasserre, J.B. (1988). A forecast horizon and a stopping rule for general Markov decision processes, *J. Math. Anal. Appl.* **132**, 388–400.

Hernández-Lerma, O., and Marcus, S.I. (1983). Adaptive control of service in queueing systems, *Syst. Control Lett.* **3**, 283–289.

Hernández-Lerma, O., and Marcus, S.I. (1984). Optimal adaptive control of priority assignment in queueing systems, *Syst. Control Lett.* **4**, 65–72.

Hernández-Lerma, O., and Marcus, S.I. (1984a). Identification and approximation of queueing systems, *IEEE Trans. Autom. Control* **29**, 472–474.

Hernández-Lerma, O., and Marcus, S.I. (1985). Adaptive control of discounted Markov decision chains, *J. Optim. Theory Appl.* **46**, 227–235.

Hernández-Lerma, O., and Marcus, S.I. (1987). Adaptive control of Markov processes with incomplete state information and unknown parameters, *J. Optim. Theory Appl.* **52**, 227–241.

Hernández-Lerma, O., and Marcus, S.I. (1987a). Adaptive policies for discrete-time stochastic control systems with unknown disturbance distribution, *Syst. Control Lett.* **9**, 307–315.

Hernández-Lerma, O., and Marcus, S.I. (1989). Nonparametric adaptive control of discrete-time partially observable stochastic systems, *J. Math. Anal. Appl.* **137** (to appear).

Hernández-Lerma, O., and Marcus, S.I. (1989a). Discretization procedures for adaptive Markov control processes, *J. Math. Anal. Appl.* **137** (to appear).

Heyman, D.P., and Sobel, M.J. (1984). *Stochastic Models in Operations Research, Vol. II: Stochastic Optimization,* McGraw-Hill, New York.

Hijab, O. (1987). *Stabilization of Control Systems,* Springer-Verlag, New York.

Himmelberg, C.J., Parthasarathy, T., and Van Vleck, F.S. (1976). Optimal plans for dynamic programming problems, *Math. Oper. Res.* **1**, 390–394.

Hinderer, K. (1970). Foundations of Non-Stationary Dynamic Programming with Discrete Time Parameter, *Lecture Notes Oper. Res.* **33**, Springer-Verlag.

Hinderer, K. (1982). On the structure of solutions of stochastic dynamic programs, Proc. 7th Conf. on Probab. Theory (Brasov, Romania), 173–182.

Hinderer, K., and Hübner, G. (1977). Recent results on finite-stage stochastic dynamic programs, Preprint, 41st Session of the International Statistical Institute, New Delhi.

Hopp, W.J., Bean, J.C., and Smith, R.L. (1986). A new optimality criterion for non-homogeneous Markov decision processes, Preprint, Dept. of Industrial Eng. and Manage. Sci., Northwestern University.

Hordijk, A., Schweitzer, P.J., and Tijms, H. (1975). The asymptotic behavior of the minimal total expected cost for the denumerable state Markov decision model, *J. Appl. Probab.* **12**, 298–305.

Hordijk, A., and Tijms, H. (1975). A modified form of the iterative method in dynamic programming, *Ann. Statist.* **3**, 203–208.

Huber, P.J. (1967). The behavior of maximum likelihood estimates under non-standard conditions, Proc. 5th Berkeley Symp. on Math. Statist. and Probab., Vol. 1, 221–233.

Hübner, G. (1977). On the fixed points of the optimal reward operator in stochastic dynamic programming with discount factor greater than one. *Zeit. Angew. Math. Mech.* **57**, 477–480.

Iosifescu, M. (1972). On two recent papers on ergodicity in nonhomogeneous Markov chains, *Ann. Math. Statist.* **43**, 1732–1736.

Ivanov, A.V., and Kozlov, O.M. (1981). On consistency of minimum contrast estimators in the case of nonidentically distributed observations, *Theory Probab. Math. Statist.* **23**, 63–72.

Jaquette, D.L. (1972). Mathematical models for controlling growing biological populations: a survey, *Oper. Res.* **20**, 1142–1151.

Kakumanu, P. (1977). Relation between continuous and discrete time markovian decision problems, *Naval Res. Log. Quart.* **24**, 431–439.

Klimko, L.A., and Nelson, P.T. (1978). On conditional least squares estimation for stochastic processes, *Ann. Statist.* **6**, 629–642.

Kolonko, M. (1982). Strongly consistent estimation in a controlled Markov renewal model, *J. Appl. Probab.* **19**, 532–545.

Kolonko, M. (1982a). The average-optimal adaptive control of a Markov renewal model in presence of an unknown parameter, *Math. Operationsforsch. u. Statist.* **13**, Serie Optim., 567–591.

Kolonko, M. (1983). Bounds for the regret loss in dynamic programming under adaptive control, *Zeit. Oper. Res.* **27**, 17–37.

Kumar, P.R. (1985). A survey of some results in stochastic adaptive control, *SIAM J. Control Optim.* **23**, 329–380.

Kumar, P.R., and Varaiya, P. (1986). *Stochastic Systems: Estimation, Identification, and Adaptive Control,* Prentice-Hall, Englewood Cliffs, N.J.

Kurano, M. (1972). Discrete-time markovian decision processes with an unknown parameter—average return criterion, *J. Oper. Res. Soc. Japan* **15**, 67–76.

Kurano, M. (1983). Adaptive policies in Markov decision processes with uncertain transition matrices, *J. Inform. Optim. Sci.* **4**, 21–40.

Kurano, M. (1985). Average-optimal adaptive policies in semi-Markov decision processes including an unknown parameter, *J. Oper. Res. Soc. Japan* **28**, 252–266.

Kurano, M. (1986). Markov decision processes with a Borel measurable cost function: the average reward case, *Math. Oper. Res.* **11**, 309–320.

Kurano, M. (1987). Learning algorithms for Markov decision processes, *J. Appl. Probab.* **24**, 270–276.

Kushner, H. (1971). *Introduction to Stochastic Control,* Holt, Rinehart and Winston, New York.

Langen, H.J. (1981). Convergence of dynamic programming models, *Math. Oper. Res.* **6**, 493–512.

Lefevre, C. (1981). Optimal control of a birth and death epidemic process, *Oper. Res.* **29**, 971–982.

Lembersky, M.R. (1978). The application of Markov decision processes to forestry management. In: Puterman (1978), pp. 207–219.

Ljung, L. (1981). Analysis of a general recursive prediction error identification algorithm, *Automatica* **17**, 89–99.

Loeve, M. (1978). *Probability Theory II,* 4th edition, Springer-Verlag, Berlin.

Loges, W. (1986). Estimation of parameters for Hilbert space-valued partially observable stochastic processes, *J. Multivar. Anal.* **20**, 161–174.

Lovejoy, W.S. (1984). Decision problems in marine fisheries, Proc. 23rd IEEE Conf. on Decision and Control (Las Vegas, Dec. 1984) **3**, 1671–1674.

Ludwig, D., and Walters, C.J. (1982). Optimal harvesting with imprecise parameter estimates, *Ecological Modelling* **14**, 273–292.

Luenberger, D.G. (1969). *Optimization by Vector Space Methods,* Wiley, New York.

Maigret, N. (1979). Statistique des chaines controlées felleriennes, *Asterisque* **68**, 143–169.

Mandl, P. (1974). Estimation and control in Markov chains, *Adv. Appl. Probab.* **6**, 40–60.

Mandl, P. (1979). On the adaptive control of countable Markov chains. In: *Probability Theory,* Banach Centre Publications, Vol. 5, PWB-Polish Scientific Publishers, Warsaw, 159–173.

Mandl, P., and Hübner, G. (1985). Transient phenomena and self-optimizing control of Markov chains, *Acta Universitatis Carolinae-Math. et Phys.* **26**, 35–51.

Mine, H., and Osaki, S. (1970). *Markovian Decision Processes,* American Elsevier, New York.

Mine, H., and Tabata, Y. (1970). Linear programming and markovian decision problems, *J. Appl. Probab.* **7**, 657–666.

Monahan, G.E. (1982). A survey of partially observable Markov decision processes, *Manage. Sci.* **28**, 1–16.

Morton, R. (1973). Optimal control of stationary Markov processes, *Stoch. Proc. Appl.* **1**, 237–249.

Nummelin, E., and Tuominen, P. (1982). Geometric ergodicity of Harris recurrent Markov chains with applications to renewal theory, *Stoch. Proc. Appl.* **12**, 187–202.

Palm, W.J. (1977). Stochastic control problems in fishery management. In: *Differential Games and Control Theory* II (edited by E.O. Roxin, P.T. Liu, and R.L. Sternberg), Marcel Dekker, New York, 65–82.

Parthasarathy, K.R. (1967). *Probability Measures on Metric Spaces,* Academic Press, New York.

Pfanzagl, J. (1969). On the measurability and consistency of minimum contrast estimates, *Metrika* **14**, 249–272.

Prakasa Rao, B.L.S. (1983). *Nonparametric Functional Estimation*, Academic Press, New York.

Puterman, M.L. (1978), editor. *Dynamic Programming and Its Applications*, Academic Press, New York.

Rhenius, D. (1974). Incomplete information in markovian decision models, *Ann. Statist.* **2**, 1327–1334.

Rieder, U. (1975). Bayesian dynamic programming, *Adv. Appl. Probab.* **7**, 330–348.

Rieder, U. (1979). On non-discounted dynamic programming with arbitrary state space, University of Ulm, FRG.

Rieder, U., and Wagner, H. (1986). Regret-Abschätzungen für stochastiche Kontrollmodele unter Unsicherheit, University of Ulm, FRG.

Riordon, J.S. (1969). An adaptive automaton controller for discrete-time Markov processes, *Automatica* **5**, 721–730.

Ross, S.M. (1968). Arbitrary state markovian decision processes, *Ann. Math. Statist.* **39**, 2118–2122.

Ross, S.M. (1970). *Applied Probability Models with Optimization Applications*, Holden-Day, San Francisco.

Ross, S.M. (1971). On the nonexistence of ϵ-optimal randomized stationary policies in average cost Markov decision models, *Ann. Math. Statis.* **42**, 1767–1768.

Ross, S.M. (1983). *Introduction to Stochastic Dynamic Programming*, Academic Press, New York.

Royden, H.L. (1968). *Real Analysis*, 2nd edition, Macmillan, New York.

Sawaragi, Y., and Yoshikawa, T. (1970). Discrete-time markovian decision processes with incomplete state observation, *Ann. Math. Statist.* **41**, 78–86.

Schäl, M. (1975). Conditions for optimality in dynamic programming and for the limit of n-stage optimal policies to be optimal, *Z. Wahrs. verw. Geb.* **32**, 179–196.

Schäl, M. (1979). On dynamic programming and statistical decision theory, *Ann. Statist.* **7**, 432–445.

Schäl, M. (1981). Estimation and control in discounted stochastic dynamic programming. Preprint No. 428, Institute for Applied Math., University of Bonn; a slightly different version appeared in *Stochastics* **20** (1987), 51–71.

Schuster, E., and Yakowitz, S. (1979). Contributions to the theory of nonparametric regression, with application to system identification, *Ann. Statist.* **7**, 139–149.

Schweitzer, P.J. (1971). Iterative solution of the functional equations of undiscounted Markov renewal programming, *J. Math. Anal. Appl.* **34**, 495–501.

Serfozo, R. (1979). An equivalence between continuous and discrete time Markov decision processes, *Oper. Res.* **27**, 616–620.

Stidham, S. (1981). On the convergence of successive approximations in dynamic programming with non-zero terminal reward, *Zeit. Oper. Res.* **25**, 57–77.

Strauch, R.E. (1966). Negative dynamic programming, *Ann. Math. Statist.* **37**, 871–890.

Striebel, C. (1975). Optimal Control of Discrete Time Stochastic Systems, *Lecture Notes Econ. Math. Syst.* **110**, Springer-Verlag, Berlin.

Thomas, L.C., and Stengos, D. (1984). Finite state approximations algorithms for average cost denumerable Markov decision processes, OR *Spektrum* **6**.

Tijms, H.C. (1975). On dynamic programming with arbitrary state space, compact action space and the average reward as criterion, Report BW 55/75, Mathematisch Centrum, Amsterdam.

Ueno, T. (1957). Some limit theorems for temporally discrete Markov processes, J. Fac. Science, University of Tokyo **7**, Sect. 1, Part 4, 449–462.

Van Hee, K.M. (1978). Bayesian Control of Markov Chains, *Mathematical Centre Tract* **95**, Mathematisch Centrum, Amsterdam.

Wakuta, K. (1981). Semi-Markov decision processes with incomplete state observation: average cost criterion, *J. Oper. Res. Soc. Japan* **24**, 95–108.

Wakuta, K. (1982). Semi-Markov decision processes with incomplete state observation: discounted cost criterion, *J. Oper. Res. Soc. Japan* **25**, 351–362.

Wakuta, K. (1987). The Bellman's principle of optimality in the discounted dynamic programming, *J. Math. Anal. Appl.* **125**, 213–217.

Waldmann, K.H. (1980). On the optimality of (z, Z)- order policies in adaptive inventory control, *Zeit. Oper. Res.* **24**, 61–67.

Waldmann, K.H. (1984). Inventory control in randomly varying environments, *SIAM J. Appl. Math.* **44**, 657–666.

Walters, C.J. (1978). Some dynamic programming applications in fisheries management. In: Puterman (1978), 233–246.

Walters, C.J., and Hilborn, R. (1976). Adaptive control of fishing systems, *J. Fish. Res. Board Canada* **33**, 145–159.

Wang, R.C. (1976). Computing optimal quality control policies—two actions, *J. Appl. Probab.* **13**, 826–832.

White, C.C. (1978). Optimal inspection and repair of a production process subject to deterioration, *J. Oper. Res. Soc.* **29**, 235–243.

White, C.C., and Eldeib, H.K. (1987). Markov decision processes with imprecise transition probabilities. Preprint, Dept. of Systems Eng., University of Virginia, Charlottesville, Virginia 22901.

White, D.J. (1963). Dynamic programming, Markov chains, and the method of successive approximations, *J. Math. Anal. Appl.* **6**, 373–376.

White, D.J. (1980). Finite state approximations for denumerable state infinite horizon discounted Markov decision processes, *J. Math. Anal. Appl.* **74**, 292–295.

White, D.J. (1985). Real applications of Markov decision processes, *Interfaces* **15**, 73–83.

Whitt, W. (1978/79). Approximation of dynamic programs, I, II, *Math. Oper. Res.* **3**, 231–243; **4**, 179–185.

Wickwire, K.M. (1977). Mathematical models for the control of pests and infectious diseases: a survey, *Theor. Pop. Biol.* **11**, 182–238.

Wolfowitz, J. (1957). The minimum distance method, *Ann. Math. Statist.* **28**, 75–88.

Yakowitz, S. (1979). Nonparametric estimation of Markov transition functions, *Ann. Statist.* **7**, 671–679.

Yakowitz, S. (1982). Dynamic programming applications in water resources, *Water Resources Res.* **18**, 673–696.

Yakowitz, S. (1985). Nonparametric density estimation, prediction, and regression for Markov sequences, *J. Amer. Statist. Assoc.* **80**, 215–221.

Yakowitz, S., Williams, T.L., and Williams, G.D. (1976). Surveillance of several Markov targets, *IEEE Trans. Inform. Theory* **22**, 716–724.

Yushkevich, A.A. (1976). Reduction of a controlled Markov model with incomplete data to a problem with complete information in the case of Borel state and control spaces, *Theory Probab. Appl.* **21**, 153–158.

Yushkevich, A.A., and Chitashvili, R. Ya. (1982). Controlled random sequences and Markov chains, *Russian Math. Surveys* **37**, 239–274.

Author Index

The numbers in this index refer to sections and appendices.

Subject Index

The numbers in this index refer to sections and appendices.

Applied Mathematical Sciences

cont. from page ii

55. *Yosida:* Operational Calculus: A Theory of Hyperfunctions.
56. *Chang/Howes:* Nonlinear Singular Perturbation Phenomena: Theory and Applications.
57. *Reinhardt:* Analysis of Approximation Methods for Differential and Integral Equations.
58. *Dwoyer/Hussaini/Voigt (eds.):* Theoretical Approaches to Turbulence.
59. *Sanders/Verhulst:* Averaging Methods in Nonlinear Dynamical Systems.
60. *Ghil/Childress:* Topics in Geophysical Dynamics: Atmospheric Dynamics, Dynamo Theory and Climate Dynamics.
61. *Sattinger/Weaver:* Lie Groups and Algebras with Applications to Physics, Geometry, and Mechanics.
62. *LaSalle:* The Stability and Control of Discrete Processes.
63. *Grasman:* Asymptotic Methods of Relaxation Oscillations and Applications.
64. *Hsu:* Cell-to-Cell Mapping: A Method of Global Analysis for Nonlinear Systems.
65. *Rand/Armbruster:* Perturbation Methods, Bifurcation Theory and Computer Algebra.
66. *Hlaváček/Haslinger/Nečas/Lovíšek:* Solution of Variational Inequalities in Mechanics.
67. *Cercignani:* The Boltzmann Equation and Its Applications.
68. *Temam:* Infinite Dimensional Dynamical System in Mechanics and Physics.
69. *Golubitsky/Stewart/Schaeffer:* Singularities and Groups in Bifurcation Theory, Vol. II.
70. *Constantin/Foias/Nicolaenko/Temam:* Integral Manifolds and Inertial Manifolds for Dissipative Partial Differential Equations.
71. *Catlin:* Estimation, Control, and the Discrete Kalman Filter.
72. *Lochak/Meunier:* Multiphase Averaging for Classical Systems.
73. *Wiggins:* Global Bifurcations and Chaos.
74. *Mawhin/Willem:* Critical Point Theory and Hamiltonian Systems.
75. *Abraham/Marsden/Ratiu:* Manifolds, Tensor Analysis, and Applications, 2nd ed.
76. *Lagerstrom:* Matched Asymptotic Expansions: Ideas and Techniques.
77. *Aldous:* Probability Approximations via the Poisson Clumping Heuristic.
78. *Dacorogna:* Direct Methods in the Calculus of Variations.
79. *Hernández-Lerma:* Adaptive Markov Control Processes.
80. *Lawden:* Elliptic Functions and Applications.
81. *Bluman/Kumei:* Symmetries and Differential Equations.